THE
WEST RIDING
MINERS
AND
SIR WILLIAM
GARFORTH

THE WEST RIDING MINERS AND SIR WILLIAM GARFORTH

BRYAN FRASER

The History Press

First published 2009

The History Press
The Mill, Brimscombe Port
Stroud, Gloucestershire, GL5 2QG
www.thehistorypress.co.uk

British Library Cataloguing in Publication Data.
A catalogue record for this book is available from the British Library.

ISBN 978 0 7524 4991 3

Typesetting and origination by The History Press
Printed in Great Britain

CONTENTS

INTRODUCTION

This is the story of the West Riding Colliery and its mining village at Altofts, which is now part of the town of Normanton in the West Riding of Yorkshire. The joint efforts of William Edward Garforth and the young Mary Eager to improve the lot of the working man resulted in the saving of many thousands of lives in the mining industry. The pair were determined to ensure that the village of Altofts' mining disaster would not be in vain.

This book portrays the meeting of William and Mary, how the explosion occurred, William's patient scientific approach to finding the cause and how, withstanding abuse and ridicule, he fought to convince not only the colliery owners, but over 1 million men who worked in them, of the importance of that discovery. It also demonstrates that a scientific approach had to be made to improve the health, safety and production in Britain's mines.

As well as the story of William and Mary, the book also illustrates why the village of Altofts, for a short period in 1908, became the centre of the mining world, and why thousands of windows were blown out as engineers from many different countries came to see what was happening at the mine.

We also follow the lives of the miners inspired by William. In the world's first Mines Rescue Station, these men gave their free time to train fellow colliers and answer their calls for help. They, and Sir William Garforth, eventually received international acclaim and forced the government to change the laws regarding mining.

Every major incident in this book is recorded, and every little part of those major incidents were passed on to me by children of the men involved. I have used my knowledge of the mining industry and the Altofts area to put the two together. Very occasionally, when I did not know the name, I used the name of local families or miners, who could have been there (with the exception of Jack and Dorothy Dyson) to help me paint word pictures. Jack Dyson, my stepfather, spent his working life at the nearby St Johns Newland Colliery. As a colliery official he was fiercely loyal to the management and men. Dorothy is my mother.

This book is written in memory of the millions of miners, including my father, who have given their lives for the industry. Without the people of Normanton's and Altofts' memories, and the notes Mr Garforth made in his small, neat handwriting on his personal copies of the *Institute of Mining Engineers'* journal, this book would not have been possible and future generations would not be aware that what happened in this small place eventually saved thousands of lives.

Thank you for sharing your memories.

one

THE MEETING OF
MR W.E. GARFORTH AND
MISS M. EAGER

William Edward Garforth was born on 30 December 1845, the second son of William Garforth JP of Dukinfield, owner of the Dukinfield Iron Works and the Lordsfield Colliery near Aston-under-Lyne. The area was booming. It supplied enormous quantities of coal to Manchester by canal. The nearby Aston-under-Lyne had grown from a population of 23,000 in 1816 to 80,000 in 1871. In 1842 Dukinfield collieries employed 150 men; by 1875 six collieries with 1,120 men produced 336.440 tons of coal.

Conditions for the workers were bad. Miners worked eleven or twelve hours per day, and sometimes up to thirty-six hours in ineffectively ventilated pillar-and-stall headings, working by the light of a candle. In 1866, at Oaks Colliery near Barnsley, 361 men and boys were killed in an underground explosion. There were many other disasters that year. At Victoria Colliery, Cheshire, thirty-eight were killed. At Talk o' the Hill Colliery, Staffordshire, ninety-one were killed. At High Brooks Colliery, Lancashire, thirty were killed. At Victoria Colliery, Cheshire, thirty-eight were killed. At Pelton Colliery, County Durham, twenty-four were killed.

Year after year, disasters continued.

Miners' homes were rows of houses packed together, mud roads and pavements which in wet weather were in a deplorable state, without proper sanitation and under a permanently smoky haze. Conditions were so harsh that from time to time the colliers rebelled; in 1858 riots took place in Dukinfield and eleven colliers were committed to Chester Assizes.

Now that we understand the working conditions, let us imagine how William met his future wife.

23 JANUARY 1868

'Fire!'

Twenty rounds were discharged simultaneously over the heads of the rioters. As the smoke from the rifles cleared and the sound of breaking glass diminished, silent heads looked out of windows and doors towards the soldiers.

'Carry out the unload then advance half your squad and seize the ring leaders Sergeant,' Lieutenant Garforth's calm voice stood out, as his short thick fingers lightly tapped the red stripe on his trouser leg.

'Yes sir,' Sgt Fraser's waxed moustache quivered as he saluted then executed a smart about turn and faced the forty men in two ranks of the 23rd Lancashire Rifles.

'Front rank will unload – Unload!' Sgt Fraser shouted the command. The crowd watched as though mesmerised, as twenty rifle bolts rattled back and forward.

'Rear rank will lay their weapons on the ground and advance at the double with me to arrest the rioters.' The soldiers prepared for the order, 'Lay down your weapons!' Twenty rifles were gently laid on the snow-covered ground.

'Advance!' The second line of soldiers broke through the first line and ran across the muddy square towards the rioting group. The few people near the door of the lamp room of the Astley Deep Colliery office buildings became a struggling mass of men moving back into the dark, away from the glow of burning caused by the smashing of oil lamps. The last few to emerge from the colliery office were grabbed and soldiers rained blows on them until they were face down in the mud. They had been called out of their barracks at nearby Ashton-Under-Lyne on a cold January day, and marched to Dukinfield – they were not happy.

A small team of men in shirt sleeves dragged fire carts into the square, one containing a mobile pump and hoses, the other ladders and buckets. A hose was quickly run out to the nearby stream and connected to the wooden pump. Preparations finished, they turned and looked at the young officer.

'Go!' At Lt Garforth's command the Dukinfield Iron Works fire brigade moved into action and quickly dragged the cart forward to the burning buildings. Two men then stripped to the waist and mounted the pump cart. Their see-saw action on the large pump handles sent water via the hose onto the fire, and so it was quickly extinguished.

Twenty-three-year-old Lt William Edward Garforth JP watched as the prisoners were secured and then brought back to face him. A couple of them bled from the mouths, and others had bruising around their faces.

'Did any of the prisoners give any reason for the riot, Sergeant?'

'One mentioned a William Murphy. That was all, Sir!'

'Okay, place the prisoners in the lock-up on Pickford Lane and mount an all night guard. I will be at home if you require me.'

'Yes Sir! The cold and damp in there will soon calm them down Sir!' Sgt Fraser said as he sprang to attention.

William's long strides soon took him to the well-lit windows of the family home. As he started to close the door he caught sight of a dark-haired girl running up the steps. She was dressed in dark brown town clothes, with her sleeves rolled up and carrying a wickerwork basket. Breathing heavily, she explained, 'I'm Mary Eager, Canon Eager's daughter.'

William looked at her closely; he had heard much of Canon Eager and his family's work amongst the poor of nearby Audenshaw. Her blush as he continued to stare at her reminded him of his manners.

'How can I help?' he asked.

'Are you the officer who's just arrested those men?' She went on, without waiting for him to answer, 'They only wanted food so they can feed their families.' She stopped and looked at him.

'They smashed up our colliery lamp room,' he said.

'They were desperate; I've just come from Ian Partington's home, one of the men you have arrested,' she said as if this was a good reason. As he continued to look, he thought,

'She must only be about fourteen years old, and yet so intense.' He then led her towards a chair, and waited for her to continue.

'Mrs Partington is eight months' pregnant.' Her shoulders slumped as she sat down. 'They have four other children and no money.'

'Why does he not sign on for poor relief?' he asked, kneeling down beside her.

'Poor relief!' Her shoulders straightened. 'His weekly wage was 13s. He could hardly manage on that, now they have reduced the relief money to 3s per week.' She stood up.

As William stood up he compared his polished boots with the mud around the hem of her long skirt. 'Show me where they live,' he said.

William reached for his cloak then followed her out into the dark night. As they made their way past the stark steel cage headgear of the colliery she stopped and turned to him.

'Your pit's not working, therefore no wages.' Without waiting for an explanation, she marched down a street of small back-to-back terraced houses which surrounded the colliery. Gaunt white faces from dark doors and windows watched their progress, some of the faces breaking into prolonged coughing which was only eased by the spitting of blood into the muddy street. Within minutes he could see the flames from the few furnaces which were being stoked with coal at the family iron works. Shadow images of small, thin bodies momentarily showed as molten pig iron poured from their vessels, hissing and sparking as the iron hit cold containers, causing the children to jump as tiny bubbles exploded and showered them. Once again she stopped, looked at one small child as he ran past, then at him.

'The few pence they earn for running errands is now the only source of income for many families.' He didn't answer, so she moved on. At the corner of Town Lane a gang of men were talking in the moonlight; on seeing William they all turned and started moving towards them, waving their fists and muttering.

'He's a Garforth,' he heard, and stopped.

William stopped, Mary carried on. As the moonlight shone on her face they spluttered their apologies, and he hurried to catch up. He was lost soon after they left Cuckoo Square, as one muddy street led to another. Pale dirty faces stood out as they passed dark windows. A blurred white image was the only warning as the contents of a chamber pot were thrown out.

Eventually Mary knocked on a door, 'Come in,'

William looked around and, as his eyes adjusted to the dark, noticed that the empty fireplace, a bare table and a pile of rags were all the furniture the room possessed. Four young children dressed in dirty clothes were shivering, huddled up to a heavily pregnant, scrawny women who lay on those rags.

'Oh Miss Eager, what'll happen to him?' Sarah Partington asked, trying to rise before dropping back as a rasping cough tore through her body.

Mary turned to him, her eyes pleading, then dropped to the floor near Mrs Partington. The Lieutenant just stood and looked.

'I will see what I can do,' she said as she produced a loaf of bread from her basket, broke it up and gave some to each of the children and the woman. They ate quickly without allowing one crumb to fall to the floor.

'Please come outside.' She took his arm and once the door had shut behind her, said, 'Two of those children are not hers: their father, her cousin, was killed by a fall of rock underground. On hearing the news, his wife had a miscarriage and died.'

'Can't her family help?'

'Her father, Jerimiah, started work in the pits at six years old. His future wife was a little older when she started work underground.' She looked up at him, 'She died giving birth to Sarah. Her father was killed by a fall of rock underground ten years ago.'

'I didn't know.' No longer did she seem so young. He felt helpless, 'I will do all I can.'

No doubt she gently touched his arm like an older sister, and disappeared back into the house. It would be many weeks before the image of Mary Eager cleared from William's mind.

The Civil War in America dried up the supply of iron ore. With no need for the furnaces, the collieries had shut down. On Easter Monday 1869, the puritan priest William Murphy, a fiery orator, stirred up trouble. He hated the Catholics and led a crowd of over 300 who smashed up their church. Murphy even shattered the cross after showing it to the crowd.

A year passed, trade picked up, and the troubles died down. William became manager of Ashley Deep Colliery.

3 MARCH 1870

'David,' William Garforth paused to take a deep breath. After receiving the message he had run in the crisp morning frost to the Ashley Deep pit colliery offices, 'David Holmes, what's happened?'

'There's bin an explosion underground, Mr Garforth. Me, Abraham Elce, Matthias Stead and Henry Buxton heard it and went to try an' help.' He started coughing so William moved to the piece of wood he was being carried on, and sat him up.

Breathing easier, David carried on, 'Me an' about 200 more were lucky, fire died out quickly, but we were all unconscious when they got to us.'

'Anybody badly hurt?' William asked as he laid David back down.

'Nine dead,' said one of the carriers over his coal-blackened bare shoulder as they moved away.

'What happened?' William addressed one of his colliery officials

'These are the statements we have collected so far.' Mr Ray, the under-manager, handed over the notes, then added, 'My comments are on the bottom.'

'Thank you. Have the families been informed?'

'I've asked Mr Bidwell from the general store to attend to their immediate needs.'

'They are to be charged to me.'

'Certainly, Mr Garforth.'

'Nobody moves anything in the area of the explosion until we have examined it.' William glanced quickly through the notes, 'I should be ready to go down in 30 minutes'

'Right, Mr Garforth,' said Mr Ray as he turned towards the door, and William sat down already engrossed in the reports.

The evening sun was just disappearing as the cage rode up the shaft following the inspection. After a short discussion everything was recorded.

'That will be all for today, Mr Ray. I will let you know what changes we are to bring in to try and improve safety when I've studied everything,' said Mr Garforth.

In the months that followed, William read all the expert opinion available, and introduced new safer methods of working into the colliery, sometimes against the wishes of the colliers who seemed to fear anything new. These included the 'Longwall' system of working the coalface, instead of the traditional 'Stall and Pillar' system. The 'Longwall' system improved the flow of ventilation along the face, but was more costly to get into production. He also issued stricter instructions on how far men could work on the face without wooden supports being in place.

In March 1871, as a member of the local school board, William E. Garforth stood silently as the wife of the board's chairman, Canon Thomas Eager, was laid to rest.

Afterwards he was likely to have been formally introduced to Canon Eager's dark-haired daughter Mary, and no doubt remembered their first meeting three years previously.

William and Mary continued to meet officially and socially, as they both worked to improve conditions for the workers. Eventually Canon Thomas Eager became Chaplain of the 23rd Lancashire Rifles.

14 APRIL 1874

'Mr Garforth?' William Edward was shaken by Josiah the groom, 'There's been an explosion at Astley Pit.'

William, who had been at a mining engineers' meeting, jumped out of bed, dressed quickly, then knocked on another door in the Manchester Hotel. A fellow mining engineer opened it.

'Please explain to the meeting we've had an explosion at one of my father's pits,' William said, then ran off without waiting for a reply. On arriving home, he donned his working clothes and was soon standing with his father, who said, 'Go, do all you can, William.'

At the sound of a horse, Mr Ray, the colliery under-manager, opened the office door. 'Men were working to strengthen a roof that had fallen, when more roof along the roadway collapsed, trapping them and also releasing some gas,' he informed Mr Garforth as the latter dismounted.

'How do you know that?' William asked quietly.

'David Holmes was having his snap nearby. Let him explain.' He pointed to David who was sat drinking a mug of tea.

'What happened, David?'

David, with a full mouth of tea, spluttered when Mr Garforth addressed him then said, 'I was eating me snap when suddenly there was a terrific roar followed by a gust of wind which blew out all us candles.' He put his mug down, 'We all started to walk towards t' explosion. I fell over a dead horse in t' dark then we came to a fall of rock and we smelt gas.' He stopped as the smell came back to him then continued, 'I kept me mouth shut and held me nose. Some ran shouting "gas" – they soon collapsed.' William waited whilst David visualised the scene, 'Most of us laid down below gas an' I remember thinking I'm going dee this time.'

All voices in the office had now stopped as they listened. 'We laid there hour after hour, nobody dared to speak, all thinking rest were deed.' He looked up at William and continued, 'Then we heard sound o' picks on other side o' fall. We still laid there. It were a long time till we saw candle light.'

'It took them all night to get through the fall,' a voice said behind William.

'I tried t' waken man laying next to me. He were deed,' David said as he looked round. 'I checked all o' them that were left. They were deed, I were last one out.'

'Thank you David, you've been lucky twice.' William turned to Mr Ray and added, 'Now tell me what you know, let's make sure there is never a third.'

'Fire started in old workings. We've been trying to stifle it out by stopping air getting to it, but some gas must have been leaking out of them.' Mr Ray paused and breathed deeply, 'We will have to seal off the entire pit if we want to save it.'

'How many men underground?' asked William.

'Fifty-four. We've tried to reach them but rescuers were overcome – thank goodness they had a rope tied to them so we managed to pull them back, or there would have been more.'

'Who is still down?' William turned at the sound of a woman's voice to find Mary Eager there. 'Give Mary a list.' He touched her hand briefly then set off towards the shaft from which a thin stream of smoke rose.

When he reached the iron cage headgear he waited while the cage came to a halt. He could smell burning and an official staggered out, coughing. On seeing William, he said, 'We've tried every underground road, the fire's spreading so rapidly we can't get through.'

William looked down the shaft; the smoke and smell was getting stronger. He turned round and addressed the officials behind him. 'Seal it.'

'Is there nothing we can do?' asked Mary, who had followed him.

'It's too dangerous, and the longer we let it burn the longer we will be in getting it back into production.' He turned and took her hands as tears ran from her eyes, 'It's bad enough to lose their men, but we must get the pit back into production or they will all starve.'

'Will you visit the families with me?' Mary asked.

'Yes, and I'll do all I can,' he replied.

The newspapers of 1874 read:

FIFTY FOUR KILLED IN GAS EXPLOSION AT ASTLEY DEEP COLLIERY.

In 1875, William E. Garforth was one of the first to pass his colliery manager's certificate (the Coal Mines Regulation Act of 1872 ensured that for the first time mine managers must hold a certificate of competence). That same year he cut the first sod as a director of what was to become the deepest mine in Great Britain, the Ashton Moss Colliery.

The years passed, and in 1877 William saved Ashton Corporation £12,000 by designing their sewerage works. He managed this by ensuring the works would not be damaged by subsidence, and with a concrete bed underneath, the land could still be mined.

Colliery disasters that year included 207 killed at the Blantyre No. 2 pit in Lancashire and thirty-six killed at the Pemberton Colliery. The Blantyre disaster, Scotland's worst mining accident, was caused by a shot being fired whilst gas, also known as fire damp, was about. William had given, and continued to give, talks to owners and miners on the dangers of this practice and how it could be reduced by using compressed air instead of gunpowder. None wished to know, but explosions continued in the mines.

In March 1878, forty-three were killed by an underground explosion at Unity Brook Colliery, Lancashire. In June, 189 were killed in an underground explosion at Haydock

The West Riding Colliery.

Colliery Lancashire, and in September, 268 were killed in a fire damp explosion at the Prince of Wales Pit in Monmouthshire.

William was convinced that a colliery could be made safer by better working practices, for example the introduction of machines to cut the coal, which would improve production and ventilation. On this theme he gave many talks to various national and local associations. The owners said new machines were expensive, and the miners said they did away with men's jobs. These talks were also published in the *Institute of Mining Engineers'* journal.

On 3 July 1879 William was offered the job of manager of the Pope & Pearsons, West Riding Colliery at Altofts, near Normanton in the West Riding of Yorkshire, and was paid a handsome salary of £500 a year. It was one of the most progressive collieries in the country.

two

1879-1886
INTRODUCTION TO ALTOFTS

In 1838 *Whites Directory* had described Altofts:

> It is a pleasant place, seated on the south side of the River Calder four miles from Wakefield and it has a town ship of 437 inhabitants. It his to be the junction of the York and North Midlands Railway called the West Riding Junction.

It was at that junction that Pope & Pearsons had sunk their colliery in 1851. Some years later J.S. Fletcher's *History of Yorkshire* described the area:

> An uninviting country, on all sides are evidence of industry and unremitting toil. The area is thick with coal dust and the sky, even on the brightest day seems to be obscured by the haze of continually rising smoke, and the vegetation such as it is, is covered by a sooty mantle while the trees are stunted and gnarled as if some evil wood spirit had caused a blight to fall.

William E. Garforth moved into Halesfield House in Altofts, and within months was involved in the Altofts Council. The following year he was elected Chairman, and held the position until he declined it in 1921. He also made an immediate impact at the colliery.

Let us look at one incident that took place:

'We should set roof supports, you must be abart 8 yards in now,' George Spur said as his shovel slid under the overhanging coal and removed some of the rocks that Thomas Calvert had pushed behind him.

'It'll be all rate, another half a yard o' coal an' it'll make a good wage packet for us this week,' said Thomas as once again he twisted his shoulders to attack the rock under the coal with his pick.

'Blast!' Thomas, momentarily distracted by George, had swung his pick too far back and in the confined space had knocked his candle down. Its light flickered briefly in the 12in-high opening created by him, then went out.

George stopped shovelling, got off his knees and picked up his candle. Then, laying full length on the floor, he squeezed his way under the coal with the candle held out in front of him.

'Right, Tom.' Tom twisted his shoulders in the confined space he was laid in and stretched out his arm behind him, holding the candle.

Working the coal face. Initially the floor level material is removed.

The material under the coal is thrown away and wood prop supports set.

The distance the men worked under the coal can be seen clearly here.

The supports were removed, so that the coal dropped. The large lumps were then carefully loaded into mine cars.

Ben Pickard, a miner, was Normanton's first MP. The tradition of having a miner as MP was carried on for over 100 years; its first non-mining MP was Ed Balls.

George's movement to light the candle disturbed the roof, and small pieces of coal dropped. As he looked up, a crack appeared. He flung his candle behind him then grasped Thomas' legs and wriggled backwards.

'What the?' said Thomas, who with pick and candle in hand was being scratched on his bare chest and legs by his movement backward. The seconds seemed like an eternity as they scrambled back, then with a sharp crack and rush of air, the unsupported roof dropped. Isiah Jones rushed forward through the swirling dust.

'Me hands are trapped,' Thomas said as he heard his approach. Isiah dropped to his knees and George, who had managed to get clear, moved beside him, and worked quickly to clear Thomas's hands to pull him clear. They were just about to relight their candles when the new manager of the colliery asked,

'What's happening here?'

Isiah Jones turned and held up his safety lamp, 'Mr Garforth.'

'Yes, it's Isiah Jones is it not, are you the official in charge?'

'Mr Garforth, these two idiots were working 8 yards under the coal without roof supports.'

Mr Garforth's safety lamp shone on sweating, coal-blackened men in short trousers and clogs.

Isiah continued, 'They are both experienced men, they know that to protect their safety we have a rule that you should not extract more than 2 yards of coal without setting roof supports.'

'You could have been killed,' said Mr Garforth slowly, 'How would your families exist?' Neither answered or raised their heads. Still they did not answer.

'I will not have men putting their lives at risk. Relight your candles and go and wait in my office until I come out.'

Mr Garforth waited until the two men had started their long walk to the pit bottom then turned to Isiah, 'Now show me the rest of your coaling district.'

The men had their wages reduced and were informed that if it were not for their families they would be out of work. Both men informed the village that the new manager was a firm disciplinarian.

The Yorkshire Miners Union, whose first office was in Altofts, and later Normanton, were fighting to improve the conditions and wages of the miners. However, not all miners belonged to the unions so they were very weak. Mr Ben Pickard, the secretary, had many meetings with Mr Garforth and over the years they developed mutual respect for each other.

Even when the West Yorkshire Miners Association moved to Wakefield, Mr Pickard was still a regular visitor. He tried to get more pay for the men, even though West Riding Pit paid its men more than any other local colliery. Mr Garforth explained that to pay more, they would have to increase the price of the coal sold.

At that time the colliery paid a high proportion of the rates of the village. On accepting the offer to join the board that ran the affairs of the village, he found out they had borrowed over £3,000 to build a council chamber and burial ground, while the sanitation in the village was non-existent. People drank from the local becks which were as polluted as the river, with the result that in 1880, of 151 births, fifty died before reaching five years of age. Under his chairmanship, the Altofts Council slowly tackled the problems. They approved the laying of water mains, and agreed to create committees to deal with other problems.

In February 1880, aged seventy-four, the owner of the colliery, Mr George Pearson, died whilst visiting underground workings with Mr Garforth. At that time, Mr Garforth was experimenting with Mr Swan's discovery of incandescent or electric light to see if he could overcome its big disadvantage, that it would not detect gas.

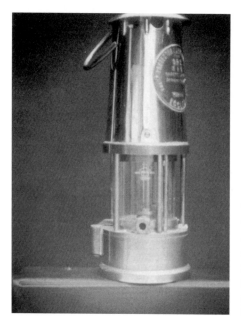

The Garforth Safety Lamp, still carried by colliery officials today, over 120 years since it was invented. Officials used a small rubber bulb to take a sample of the air underground. Then they inserted the mouth of the bulb into the hole (centre) of the safety lamp and slowly pressed the bulb. Any gas present would show by a different shape/colour on the flame.

In March 1880 the coal trade went through one of is depressions, which resulted in many families emigrating to America. As the need for coal decreased, William had to reduce men's wages by 10 per cent so he could reduce the price of coal. There were so many men out of work that the other local town, Normanton, could not pay the wages of their road workers who only received 1*s* 3*d* per day, so the roads were in a terrible muddy condition. The Salvation Army spent a great deal of time rescuing men who in their depressed drunken stupor had fallen into the deep mud. They were determined to wake them up to the realisation that drunkenness was not the answer to the depression.

William was always looking for ways to improve safety at the colliery. Any of the latest mining inventions that he though would improve safety and output were tried, including various steam and electric coal-cutters.

Let us look at how in 1882 he improved an invention. That is still in use worldwide today.

We can picture William collecting his top hat and walking to the engineering workshop.

'Mr Bramley,' Mr Garforth shouted over the sound of the steam hammer to his young colliery engineer, 'I don't like the men using candles underground.'

'Their argument is they don't know when gas is around with the safety lamps,' Mr Bramley replied, looking round and moving as sparks flew from welding equipment one of his assistants was using.

Mr Garforth waited until they both moved into the engineer's small office and the door was closed, then said, 'Yes, but that same blue flame that shows on their candles and tells them gas is around could cause that gas to explode.'

'We both know that Mr Garforth, but the men don't.'

'That's true, so we have to find some way of testing for gas with the safety lamps – I've had an idea, William Garforth produced a drawing of a safety lamp from his pocket. On the copper housing at the base of the lamp he had drawn a small hole.

'That hole is for this.' Another piece of paper was placed on his table, and showed a small bag with a copper mouth. William went on, 'We design this bag to take samples of air, and that hole has a special valve so it only works under pressure. We can control by pressing the flow of the sample over the flame on the lamp, so if it contains gas it will show on the flame.'

Sam's face lit up, 'It's brilliant. And so simple. Leave it to me.'

In September 1882 Mary Elizabeth Eager became Mrs Garforth, and soon became involved in her new community.

Let us take a look at what was happening.

'It's been such a marvellous summer holiday, I will remember 1882 for a long time,' Hannah Raybould spoke to her neighbour Sarah Holmes as they looked towards their husbands, Fred Holmes and Richard Raybould, who were stripped to the waist, helping the local farmer harvest the corn.

'Our Elizabeth and your John and James are going to be all scratched, playing where corn's bin cut. We'd better get 'em home, it's summer concert tonight in aid o' band funds,' answered Sarah.

In the West Riding Colliery School, a hall had been created by pushing wooden class dividers back against the wall. In a space at one end, local people sang or cracked jokes

Miners and their families helping local farmers.

The colliery school, also used as a community centre (closed in 1946).

Halesfield House, the Garforth family's first home.

to good hearted applause from the audience, who were crammed into the remainder of the space. Then Mrs Garforth, in her dark brown gown, moved forward to the pianoforte, and for the next twenty minutes they sat or stood in rapturous silence, whilst her small, dainty fingers danced over the keys. The cheers were instantaneous, and the people standing on the long wooden desks, which had been moved to the back of the large double classroom, stamped on them. The applause continued as they welcomed their managing director's new wife into the community with a single rose, presented by young Elizabeth Holmes.

Mr Garforth advanced to stand with his wife. She turned, looked at him then faced the audience and said, 'Thank you for inviting me. Halesfield House is now my home, but if we can help in any way please call.' In the silence she added, 'Before we finish the concert let me remind you of two things: we have a meeting at Halesfield about the colliery's Co-operative outing to Scarborough, and the children's tea party the week after. Now give all you can to the funds.'

Mary persuaded the village to raise money and convert an old maltkins into a community centre. The now-prosperous village was a must for travelling groups. One such group was 'the swinging pilgrims' who had a love feast in the local Primitive Methodist chapel. The village feast was so popular that people walked there from Castleford and Wakefield. William encouraged the formation of a band at the colliery and gave a large donation. Against learned opinion, but with the help of the local doctor, Mackenzie, he decided his colliers must learn the elements of first aid. These first-aid classes were the first in Yorkshire and maybe in Great Britain.

Mary's first child, Mary Eager Garforth, born on 2 September 1883, sadly died just one day old.

The ups and downs of the coal trade continued. Men were laid off, and the rest of the miners came out on strike. Very quickly, with no wages being paid, the prosperity of the village suffered. Whilst Mary carried on her work in the community, many families had no income. Her basket was always full as the summer came to an end, as wives with families, both cold and hungry, welcomed her into their homes behind the backs of their striking husbands.

Let us imagine the conversations at the Garforth home one day after Mary had been turned away by a wife showing signs of bruising around her face.

'William, the strike is six months old. Can nothing be done?'

'Mary, other managers would have turned them out of their houses, I'm doing all I can. Mr Pickard has been to see me today and I've explained the situation once again to him, and as soon as I am able, I will consider raising wages.'

In December, over 800 men attended a meeting in the Poplar public house and decided to withdraw their strike notice. As Christmas was so near, William immediately gave all the men one week's pay.

In the spring of 1884 the village community centre was ready to be opened.

In June Mary presented first-aid certificates in the colliery school. Their second child was due to be born the next month and, hiding what must have a worrying time, she had a smile and a friendly word for each one as she handed them the certificate, making them feel very proud.

Along with the men, with great encouragement from Mary, women also enrolled for the next course.

Between the Miners Arms and the Poplar public houses. There was a space used by the miners for mass meetings.

In July, Margaret Garforth, William and Mary's second child, was born.

Also in 1884, Mr Benjamin Pickard became the area's first MP, and was congratulated by Mr Garforth, who had refused the offer of being his Tory opposition.

Explosions still continued to kill and injure thousands of colliers. William carried on his search for safer, more profitable ways to get coal, and his research continued to be published

On 29 January 1886, Mary had her third daughter, Helen Kathleen.

three

2 OCTOBER 1886

Several small areas of light moved around in the otherwise total darkness. Wooden props supporting the roof had been removed to allow holes for the gunpowder to be drilled. The weight of millions of tons of rock above them moved onto the adjacent wooden props and they creaked and moaned under the added strain. The three holes had been packed with gunpowder, a wire inserted, then sealed.

'That connection looks fine, check other two,' Samuel Lomas said as he held his oil lamp above Martin Buxton, who had completed connecting the wires coming from the powder to slim white cables.

'They look fine. Once we've blown this lump o' rock off, roadway will be straight. All we've got to do is clean up, then we can go home,' Samuel gently ran the white cable over his hand as he walked away from the connections. The lifting of the cable and the passage of their clogs stirred up the coal dust from the floor of the Silkstone Seam.

Sam Flint and James Nicholson squatted behind a full coal tub. William Barker and John Newton moved into a shelter carved out of the rock at the side of the road. Samuel, carrying his shot-firing equipment, the ends of the cable and the sandwiches made by the youngest of his four children, moved into another similar shelter. Martin pushed his coal-blackened bare back against the rock side as Samuel entered. Then he held his oil lamp high as Samuel's large hands connected the ends to the brass terminals on the shot-firing battery. Once satisfied, he turned the handle on the equipment to build up the charge.

'Ready to fire,' he shouted, then waited a few seconds.

'Fire!' He pushed the plunger.

For a split second all was silent, then a double blast and yellow flame rocked the coal tubs. There was a brief pause then a third yellow flame came through the swirling coal dust. Sam, ears still ringing with the blast, ignored the small pieces of falling rock. He started to straighten above the level of the coal tub just as the swirling coal dust ignited then exploded into dark red flames. He was looking into a furnace; tentacles from the hungry fire moved towards him.

'Down,' he shouted, bundling his mate James to the floor. Both felt the terrific blast followed by the burning heat. Their agonising screams only lasted seconds as they were cooked alive.

In their blast-proof shelters, Samuel and William had just begun their separate forward movements when the blast and the explosion of light hit them. Gathering pace, the fire moved towards them like a fiery octopus. Martin and John's rapid backwards movement squeezed the air out of their lungs. Both reacted by pushing forward to create breathing space, then unaware they began inflating their lungs, just as arms of intense heat moved into their shelter. They were unable to even scream as probing fingers of flames travelled down their airways and over their hair, then moved down to meet others coming up.

In a short tunnel leading off the main roadway 1,000 yards from where the shots had been fired, George Colley, a plumber, was being handed a spanner by young Abraham

Davies. George asked him, 'Did thy 'ave a good fifteenth birthday?'

Abraham's answer was choked in his throat, as a blast of air flung him 30 yards. This extra distance only slightly delayed his death as the ball of fire now racing down the underground roadways reached them both.

'Amen,' sang Tom Ibbeson, who was younger than Abraham, as he sat in the main roadway rehearsing the songs for the chapel service the next day. Hearing what he thought was the noise of a horse pulling a coal tub approaching his ventilation control door, he jumped up to open it. The increased passage of air propelled the fire forward like a missile from a gun, and he caught the full blast. George Allat had just returned to his post by the steam engine at the shaft side, after watering the fifty-three ponies. They were having a well earned rest in their stables.

'What the –?' he said as they whinnied and stamped in fear as the fire reached in for them. The crackle of burning flesh replaced their agonised whinnies as they died. George placed the Bible he had been reading in his back pocket, and was moving towards them when another burst of flames found him, blasting him backwards, then running over his body like a fiery lawn mower.

'Now watch,' said Edward 'Ned' Kaye to his son, Allen. He was so pleased that Mr Fisher the colliery manager had agreed to train Allen as his assistant, 'This is how you dismantle a safety lamp.' He placed the dismantled lamp pieces on the table in the shaft bottom lamp room.

'What's that?' The screaming of the horses caused him to look out of the window. 'You stay there an' pick up pieces you dropped on t' floor,' he said, opening the door. The blast preceding the fire ball shot past him. Its initial force sent him backwards before he staggered sideways behind the cabin out of its path. He dropped face down, breaking his fall with his hands, then he heard, felt and smelled the heat as it burned his hair and clothes. Eventually he sensed the worst had passed and, using the cabin wall he got to his feet, felt the singed hair at the nape of his head and looked around. A water pipe had cracked in the intense heat, and flames darted to and fro in the steam. Then he heard the scrape as the cabin door opened; his son emerged from the steam, walking with arms outstretched like a blind man feeling his way, stumbling over rubble on the floor.

'What's up lad?' he asked.

'I'm looking for me dad, Ned Kaye.'

'You've found him,' Ned placed his arm round his son and they stumbled over the rubble to the shaft side.

'It looks like we will have to wait a bit,' said Ned looking at the cage twisted and jammed in the shaft..

Bill Burr rubbed his hands in the cold breeze. Fred Armitage the referee looked at his watch – one minute to three.

'Come on ref,' shouted the local Altofts people from the touch line nearest to the Silkstone Row houses.

'Come on ref,' the far side packed with people who had walked from nearby Normanton responded.

'Captains.' Bill tucked his shirt into his shorts, tied the lace holding them in a bow, wiped the mud from the front of his clogs on the back of his socks, and moved towards the referee. The referee placed the football on the centre spot, took a new coin out of his pocket, noted the date 1886, then said, 'Altofts call.'

With a flick the coin rose in the air: Whoosh... bang!

'What the –?' the referee said, turning in the direction of the sound.

'What the –?' Bill, along with other, said, as they turned around.

'Pit's on fire,' an arm pointed over, and 3,000 people stared into the clear blue sky, where a column of black smoke could be seen rising from a cage headgear.

'It's Silkstone Pit,' echoed cries across the field.

His shoulders dropped as he started towards the smoke, and soot began to drop on him.

'What's happened?' a young lad now pulling his mother's hand said.

'Where you all going?' asked the ref, the game forgotten, as teams joined the departing crowds.

'Don't you understand ref, it's our pit,' said Bill Burr as he grabbed his coat and ran.

'What on earth's that?' shouted Mr Fisher the colliery manager, from his study. He had been reading the latest mines safety information, when the blast caused his windows to rattle and the books on the shelves to clap like a row of applauding spectators. He rushed to the door, and found Barbara, his housemaid, pointing to smoke rising over the colliery.

'Get my underground clothes,' he shouted. Then, running towards the stables, he shouted, 'Abraham! Prepare the horse and trap.' Abraham quickly moved from where he had been watching the rising smoke, into the stable. By the time he had gone back and got the old overcoat he used for working at the colliery and returned to the door, the maids were waiting with a parcel of clothes and some sandwiches. Abraham, meanwhile, had quickly harnessed Whitey to the trap and was leading it towards the house.

'Why?' he asked himself as he climbed into the trap, 'we have always taken every precaution.' He could not give himself an answer.

As the white horse and trap approached the colliery through the falling soot, people seeing it made way, and all turned towards him. Some asked with words, the others with anxious eyes, 'Who's injured?', and most importantly, 'How badly is the pit damaged?' Without work for their men, all would starve.

Turning the corner after passing under the railway bridge, one fellow traveller shouted, 'Look!' and pointed. The large muttering crowd were also staring at a thin plume of smoke which was rising above the stark, steel structure of the Silkstone upcast shaft cage headgear. Festooned with rubbish, it stood out from the other shaft headgears which marked the entrance into the other pits in the Pope & Pearsons West Riding Collieries Group.

At the nearby Altofts station, the driver and passengers on a train watched, fascinated. 'There's over 500 men underground,' a passenger told his neighbour.

'Is that a body?' another pointed at a object that seemed to be clinging to a cross member on the shaft headgear.

'Mam! Mam!' young Ernest Hudson pulled his mother's hand, then he stopped as he sensed the tension and worry in her.

'Safety oil lamps are in use underground, men don't like them,' George Beever, an old collier, was explaining to a young woman, 'but they are supposed to be safer, candles give more light y' see, but flames can ignite any gas.' Then he noticed Mr Fisher and stepped from the crowd to hold his horse as he dismounted. 'Twenty-eight men and boys underground, Mr Fisher,' he said, then continued, 'day shift of over 400 men had just gone home.'

'Thanks George, send for Mr Garforth – he is at Leeds University,' Mr Fisher replied as he walked to the shaft side.

'Samuel, what's the position?' he asked Mr Bramley, the colliery engineer.

'Air is now flowing down the downcast shaft and up the upcast, however the upcast shaft cage is jammed at the shaft bottom.'

'Right Sam, we'll go down and have a look. Get the shaft inspection cradle, and sling a rope over that pulley.' He pointed at a small pulley, which at some time previous had been attached to the steel headgear; it appeared to be undamaged. Willing hands soon had the cradle swinging gently in the shaft.

'Okay, lower us down slowly. The signals are one rap on the water pipe means raise, two to lower, three to stop, okay?' Mr Fisher called to the team, holding one end of the rope, as he and Mr Bramley climbed into the small cradle attached to the other end. Two taps sounded on the water pipe running down the shaft and the team slowly fed out the rope, then three taps – they stopped. The short space of time until the two taps were heard seemed like hours, again three taps.

'What are they stopping for? Have they reached the bottom?' asked Martha Beevers.

'They're examining the brick stopping that seal t' worked out Warren House and Stanley Main Seam workings, as they g' down,' said George Beevers to his wife, 'if they've been blown out that means gas from them could be getting into Silkstone working at t' lower level.'

Martha knew the dangers of gas. Her father had been killed at a colliery in Staffordshire by a gas explosion. A year later her mother, hearing of houses being provided for workers in Yorkshire, had walked with her young family to Altofts. On the way her mother had stepped over the brush to cement the relationship with her new father.

Two taps. The relaxation of tension could be heard as hundreds breathed again. Again the rope was gently lowered, foot after foot went through their hands, then again three taps. Thirty minutes ticked by. Arm muscles tried to relax, as fresh volunteers joined the team. Forty-five minutes. One tap – strong arms slowly raised the cage. It was soon gently swinging at the surface. Willing hands pulled it to the side to allow them to get out.

'We heard survivors.' The crowd was quiet as Mr Fisher spoke. 'The brick stopping sealing off the Warren House and Stanley Main workings has blown out, first job those survivors in the pit bottom, then those stopping before more gas leaks out,' he dropped his voice. 'Organise something to repair those stoppings, Sam, after we have got the spare cage in place.'

'As soon as we have attached the steel rope we'll take off the damaged cage to the spare,' he answered, then turned. 'William.' This was addressed to the now managing director of the West Riding Collieries, Mr William Edward Garforth, who had just arrived.

Mary, his wife, stood with him along with his two young daughters. He kissed her then moved over and spoke quietly to Mr Fisher, 'Update me.'

'Explosion happened about 3 p.m., the day shift, all 400, had just gone home. There are twenty-eight men and boys underground, some of them are at the Silkstone pit bottom. I presume you heard the rest.'

'Yes, thanks.' He moved over to his wife as the spare cage was carried to the edge of the shaft.

'Mary, we both know what to do, but take care, we don't want to lose another,' and he looked down at her swollen stomach.

Mary held his hand tight and said, 'Yes, I am going home. We will see you when you are satisfied you have done everything possible.' With that she grasped the hand of her daughter, Margaret, then she pushed the carriage containing Helen, only eight months old, away and into the crowd.

Meanwhile the cage, attached now to the steel winding rope, was waiting to be lowered.

'Mr Fisher, Mr Buxton and just five of you deputies please accompany me,' Mr Garforth spoke in a precise manner, then entered the cage. All were dressed in suitable clothing and leather skull caps. George, the man in charge of the cage-loading bank, handed them their safety lamps and took a boarding token from each of them as they filed on to the cage; the last one in was given a ladder. George noted the time and wrote it on a nearby board, then signalled two on the bell to the engine house. The steam-winding engine driver acknowledged with two back before starting to lower the cage.

The rubble blasted up the shaft had scoured its brick lining clean. The smell and taste of burning was stronger as they approached the flickering light at the bottom of the shaft. The cage under the control of the skilled winding engine man came to rest gently on top of the other cage.

'Pass me the ladder,' William Garforth said. After collecting it, he positioned it in the narrow space between the straight side of the jammed cage and the rounded shaft walls, then climbed down.

'Glad to see you Mr Garforth.'

'Edward Kaye, what happened?'

'All I can remember is I just come out o' lamp room,' he pointed, 'when I saw this fireball coming, I dropped down behind cabin wall, and I felt the flames burning back of my hair and neck. We were lucky.'

'Anybody else around?' William asked.

'Joseph Whitaker and John Richardson are just along that road. We've not touched them they're so badly burnt. Oh, and just as you came down I thought I heard someone shouting.'

'Thank you Edward. We'll get them out first if you don't mind. I could ride the lad up with them, but I think it better if he stays with you. If John and Joseph are as bad as you say, where are they?'

'About 10 yards over there.' Edward pointed to some rubble on which a small fire still burned.

Mr Garforth looked, then turned towards the group of men assembling in the shaft bottom area. 'Mr Buxton, can you and three more come over here. Mr Fisher, take a couple and have a look round. Edward thinks he heard someone shouting a short while ago.'

'Let's shout and then listen,' said Mr Fisher, 'they might be able to hear us.'

'Anyone there?' they all shouted.

'Help, help!' a faint voice called back.

'That came from the road that links the two shafts. Right, let's go,' said Mr Fisher.

'Here,' the voice got stronger. They lifted a coal tub underneath which the voice had come from. It was Samuel Plimmer, aged seventeen.

'Well Samuel lad, y' new wife is going to be happy to see you. Just lay still and thank your stars tub fell on you, it probably saved your life.'

'Mr Fisher, am I pleased to see you! James Harris should be somewhere near. I thought I heard a moan a short while ago.'

'How y' feel Sam?'

'Okay, Mr Fisher, me legs hurt where tub fell on 'em, and I could do with a drink, me mouth's parched.'

'Have a drink of this.' He produced a small bottle out of a pocket.

'Water never tasted better, thanks.' He handed the bottle back.

'Sam, if you feel all right can you make your way to the pit bottom?'

'Mr Fisher,' Josiah Hill called, holding his lamp above a pile of twisted rail track. 'I've found him, he's trapped. Richard, give us a hand.' Hot metal was levered away as a moan came from James.

'Okay James?' asked Mr Fisher.

'I hurt all over, and my neck's sore.'

'Aye your neck's a bit burnt. Any pain? Can you walk?'

'I can feel blood coming back into me legs, give us a minute and I'll try. Please give us a hand, Mr Fisher.' As he helped him to his feet, Mr Fisher said,

'Well James lad, it was Thursday you got married wasn't it, your wife nearly had two dead husbands – you and your brother.'

'Priscilla's a grand lass, I thought so when she was me brother's wife, although some in village said it weren't right. When Bob got killed I was determined to marry her.'

'Right lad, make your way to shaft bottom. We're going to have a look round and try to put some of these fires out.'

'Mr Fisher,' Charlie Potts called, 'I think it's Walter Megson and just over there I think that's James Worthington. I can't tell, their faces and clothes are badly burned, but they're still breathing.'

'Right, you and Josiah stay with them, while I go see if I can get help.' He moved off through the burning rubble.

'Mr Garforth, we've found two lucky newlyweds, Samuel Palmer and James Harris. Both can walk. We also think we found Walter Megson and James Worthington. They're unconscious and very badly burned,' Mr Fisher addressed Mr Garforth, who was looking up the dark shaft. He turned, 'We managed to get Joseph and John out of the pit. I'm not fancying the same experience again but I'll give you a hand. What about the rest of you?' They all followed Mr Fisher.

'Mr Garforth?'

'You still here, Ned?'

'Something's been troubling me about that fireball. I've seen a gas explosion – it's got an orange tint. This one was dark red,' Edward Kaye explained.

'Thank you Ned, that's very interesting.'

It was 6 p.m. when Mr Garforth and all his party returned to the surface. All the casualties in the pit bottom had been brought out. He immediately walked over to where a first aid station had been set up.

'Bill,' he spoke to young Bill Burr, 'you never thought when you joined the first aid class that your first casualties would be me and all the top management of the pit.'

'Shall I get help Mr Garforth?'

'No Bill. I little thought when I asked Doctor McKenzie to train the men to do first aid, that I would be so thankful. The rest of the first aiders have plenty to do. Leave them.'

'Dr McKenzie is here and we've some more bad news. Ned Kaye collapsed when he got to pit top, we couldn't revive him.'

'Where is he?' William asked.

'Mr Fisher gave them permission to borrow your trap and his son Allen took him home.'

four

THE VILLAGE IN MOURNING

Now let us stand with that solemn crowd as they waited in the October darkness for news. No doubt rumour swept over them like waves.

Caroline Wilmot turned to her husband, Leonard, and asked,

'Will they shut the pit down? And what about our Gilbert? He's only just started.'

Mary and Patrick Davies had initially walked to the village from Ireland, looking for work. Now Mary asked her husband,

'Patrick, how are we going to feed James with no money coming in?' Patrick paused before replying,

'We wait and see. If I've got to walk and find work elsewhere you an' baby can stay wi' Margaret.'

Margaret and her husband, James, son Nathaniel and daughters, Margaret and Mary, were standing close by. They had become good friends since moving in next door and found out that she and Mary had come from the same district in Ireland.

'It won't be that bad Mary lass,' said James, 'if you lose your house there's a spare bed in Margaret and Mary's room now that Ellen's got a housemaid's job in Leeds.'

'How's she doing?' Mary asked.

'They're right good to her. They're letting her come home for t'afternoon on her fourteenth birthday next week,' Margaret replied.

'Aye, your James is okay – he don't work in Silkstone like my Pat. What seams do your lads Patrick, James, and Thos work in?' Mary asked.

'They all work in Stanley Main wi' their father,' Margaret replied.

A noise from near the offices started a wave of 'shush.' It silenced them and a dark shape ran forward, placed a box on the floor in front of one of the brick pillars which held a leg of the Silkstone shaft cage steel headgear. A heavily-built man walked forward and mounted the box. His round, white face and moustache was clear to all as it stood out in the dark. William Edward Garforth looked out over the sea of expectant faces and said in a precise manner,

'I have to inform you that there has been an explosion in the Silkstone seam. We do not know at this time the number of casualties or the extent of the damage.' The cage rope rumbled as it set off on its journey into the bowels of the earth. Mr Garforth paused till the sound died away. Quiet echoed across the alert, dark mass as he continued.

'All officials to my office.' In the moonlight William saw the crowd open their ranks to let them through, then closed and surged forward as he added, 'We will do all we can,' the crowd remained quiet so he continued, 'and keep you informed.'

'What did that mean, how many and who was dead?' Questions sailed across the crowd.

'If pit shuts down, how are we going to live with no wages being paid?' newly married Mrs Westwood asked her mother-in-law.

West Riding Colliery village seen from the direction of the colliery.

The village viewed from the opposite side; the Silkstone coal seam was being worked at the same time that Silkstone Row was built. The row was at one time the longest unbroken row in Europe.

This terrace fronted onto the main road. Note the only store for up to 4,000 inhabitants.

Each coal seam had its own winding headgear, to enable coal and men to reach the surface.

The management team that coordinated the recovery of the mine.

The colliery officials.

'We'll have to wait and see,' she answered.

Jack Dyson joined other serious-faced officials already standing around the large walnut table in the centre of Mr Garforth's office. Several were making notes as they looked at underground maps which littered the table. As Jack waited in the quiet flickering light provided by the oil lamps, he noticed the different emotions. Richard Raybould shook his head from side to side in utter bewilderment. Josiah Hill, with puffed-out cheeks and tightly drawn mouth, was constantly taking a deep breath – he had close relatives underground. Mr Buxton, clenched teeth and fists waving up and down, was eager to get started. Charlie Potts tapped his fingers together. Thos Smith swayed as he changed his weight from one leg to the other.

All movement ceased. William removed his top hat as his dark shape filled the doorway. He bent forward, placed both hands on the table and stated, 'The first priority is to save the men still underground.' Then, throwing his overcoat on a chair, he added, 'They are George Allat, Joseph Gill, William Barker, George Colley, Thomas Ibbeson, George Cookson, John Newton, Charles Plimmer, Thomas Oakley, James Nicholson, Martin Buxton, John Trueman, John Fox, Abraham Davies, John Hancock, Sam Flint, George Wilcock, Samuel Lomas, Henry Deakin, George Megson, Solomon Green, Joseph Whitacker, Edward Kaye, Allen Kaye, Jack Worthington, John Richardson, Samual Plimmer and James Harris.' Jack knew most of them. Mr Garforth continued,

'We have no wish to add to that list, so take extreme care.' His grey eyes bored into them as they swept around the room. In three long strides he was at the front of the table and seated in his chair.

'All will make notes of everything they do; we must learn from this, not only what caused the explosion so it never happens again,' he paused, 'but how we must be prepared for any emergency.' He let that sink in before adding, 'My clerk, Harold, will be available to help. Any questions?' The white face of Harold came up from his note taking. Then Mr Garforth looked at his tall under-manager.

'Right Mr Fisher, start organising the volunteers,' Mr Garforth swept around the table. 'I want three officials to help Mr Fisher.' As all the officials made their way towards the door, he said, 'Thank you gentlemen, but the first three, Bartholomew, Nathaniel and Jack, will suffice.' Seeing movement from the office, the crowd surged forward hoping for more news.

Mr Garforth now turned to the colliery engineer. 'Mr Bramley, organise all the equipment we may possibly need. Anything we have not got, inform me as soon as possible.'

'Okay,' said Samuel Bramley. Samuel was so deep in thought as he made notes in his small neat handwriting that he almost walked into Jack. William then turned to Harold,

'We had better get in touch with Mr Wardell the mines inspector. Send this.' As Harold dipped his pen into the ink, Mr Garforth said,

'Underground explosion in the Silkstone seam, fatalities possible.' Mr Garforth leaned over and added his signature. 'Okay, Harold, once it's sent, arrange for sandwiches and drinks. It's going to be a long hard night.' Harold's long white fingers took the telegram out into the night.

Mr Fisher stopped, and in turning he was illuminated by the light of the oil lamp hung on his belt. Now he was the only person they could see against the dark muttering background. His chest expanded and the blue coal dust-filled scar stood out. His face seemed to puff up as determination emanated from him.

THE COLLIERY EXPLOSION.—There is little new to report in connection with the explorations going on at the Silkstone Pit. The work is being pushed forward with unremitting vigour, under the direction of Mr. Garforth, the manager, but the excessive falls of roof that are met with makes progress necessarily slow. The road in the direction of Roper's Drift, where most of the bodies still unrecovered are lying, has now been cleared and made good for a distance of over a hundred yards. The ventilation has undergone no material alteration, and the districts previously restored continue in a satisfactory condition. Another of the sufferers by the explosion has succumbed to his injuries, James Worthington, of Beckbridge, having expired last Sunday. Worthington was only married two days before the explosion, his wife being the widow of a brother of his who was accidentally killed two years ago. The inquest on the deceased was opened by Major Taylor on Monday afternoon, but after evidence of identification had been given, it was adjourned until the full inquiry, which commences on the 15th instant. Worthington was interred at the Altofts Cemetery on Wednesday. In the Altofts Primitive Methodist chapel on Sunday night the Rev. P. T. Yarker "improved" the death of Mr. George Allatt, one of the killed in the explosion, preaching a most eloquent and appropriate sermon to a crowded audience. The rev. gentleman alluded at length to death from a scientific point of view, and referred to the deceased's connection with the society and the Sunday School. Appropriate hymns were sung by the choir, and Miss Hall, of Normanton, rendered a solo, "He sleeps in the valley." Contributions to the Relief Fund are coming in fairly well, as will be seen by the advertised list; but there is much distress in the neighbourhoood, several hundreds of miners and boys being out of employment. These are being relieved by collections under the auspices of branch of the Yorkshire Miners Association, but the small amounts obtained go but a short way towards maintaining the large families. Now that a way has been made from the New Haigh Moor Pit to the New Haigh Moor seam, it is anticipated that employment will be found for additional workmen. The distance between the two shafts is about a mile and a quarter. The depth of the pit is about 180 yards. Some distance below lies the Silkstone seam, to which it is ultimately intended to sink. A connection will then be made with the pit where the unfortunate calamity has happened. The difference between the depths of the Silkstone seam at both places is, however, rather considerable, and a good deal of difficulty will be caused thereby. Anent the collections for the explosion fund, a correspondent, who signs himself "Coal Miner," writes us a letter containing remarks which may be very good in themselves, but which would, in our opinion, scarcely be in good taste at the present juncture, and might possibly tend to check the flow of subscriptions. His suggestion that a permanent fund should be established for the relief of sufferers by colliery accidents generally, is, however, a good one.

This is a newspaper report from the time.

'We need three teams of between sixteen and twenty experienced men, each to work four hours. Jack,' he said, 'you're the most experienced. Will you take the first team?'

'Yes,' Jack answered.

'Bartholomew, the second,' Bartholomew acknowledged with a nod of his head. 'Nathaniel, the third.' Nathaniel's set and determined face gave its answer. 'Now', Mr Fisher went on, 'before going out and selecting your teams, quickly tell me of any equipment you might need.'

Mr Fisher handed Jack his oil lamp, then pulled out a pencil from his grey trouser pocket, and in the light of the lamp wrote what we required. The teams needed a water supply, hosepipes, shovels, safety lamps, rope, and ventilation equipment.' Once satisfied that everything had been covered, he turned to the waiting crowd.

'We need volunteers,' his authoritarian voice rang out. Men left their families and filtered forward through the crowd to form a solid mass of volunteers.

He turned and looked at his volunteers. 'Right, pick your teams. Jack first.'

'Eric Weaver, George Hopwood, Tony Wright, Bill Harrison, Adam Firth, Bill Burr,' Jack carried on calling out names he knew so well; all stepped forward. The other two teams were selected likewise.

Mr Fisher looked at his colliers, their dress of old coats, shirts, trousers tied with string just below the knees, clogs or boots, cloth caps and bottles of water: they testified that they were ready to go.

'Teams will be changed every four hours. Each man also please carry paper and pencil.'

'Why?' Bill Burr asked.

'Some of us need a rest, not like you footballers, Bill,' said George Hopewood. Actually it was his wife he was worried about – she had never been the same since they had dragged all their possessions on a cart from Cornwall.

'After we have rescued everyone underground, we must make sure this never happens again.' Mr Fisher interrupted.

'Never happens again,' seemed to echo back to him from the dark assembly of people. Mr Fisher dropped the level of his voice when the echo stopped, 'Every little detail must be recorded,' and then he waved them away.

As his gang looked at Jack, he thought over the years they had worked together as a team.

'We'll work in three groups of six, two in each group wi' shovels, two setting or checking the wood roof supports and two ont' alert.' They knew the dangers, they had all been involved in hazardous underground incidents.

'Just like we did when we got trapped by that roof fall and had to dig our way out?' asked George.

'Yes,' Jack answered, 'but this time we know that we will have good air.' They had all been gasping as the extra effort used up their limited supply of trapped oxygen. In the semi-darkness they moved into smaller groups, as he continued.

'Change round every thirty minutes, the two on alert, test f' gas every five minutes. Any questions?'

'Can't we let our wives know?' said Eric Weaver, a newly married Castleford man. They had all given him fatherly advice during his courting of Annie from the village.

'They'll know already, you can't cough in this village without everyone knowing,' commented Adam. Normally his well-built body shook when he laughed at his own jokes, but no-one laughed. To ease the tension he said out of the corner of his mouth to George, 'Yes, Eric you'd 'ave been embarrassed if tha' knew how news of your Annie woke her mother up, to get in after seeing you one night had been spread through t' village.'

'They thought it wo' knocker up, her father was half way to work before he checked his watch,' replied George, with a smile.

'Right,' Jack said, 'let's go. Pick up tackle, lamp and tokens.'

In the combined light of their lamps, they made their way from the brick lamp cabin towards the large oil lamp that marked the entrance to the Silkstone shaft.

'Be careful, we've not had time to clear all t' glass and hot metal away,' said George, the one-armed banksman, as they entered the cage and handed their boarding tokens to him.

'Thy's a grand one to say be careful,' said Bill. 'If thy'd been more careful and got out o' way when that wire rope came off pulley, thy'd ave two arms and could have given us a hand.'

'Well you can have my hand if you find it while y' down there,' answered George with a smile.

Crammed up tight in the diminishing light of the cage, they plunged down. Jack continued his mental check of faces. Looking down at Bob's blue scarred back and bald head, he said, 'Bob, thy's the most experienced. Will tha take the lead?'

They swallowed as their ears popped with the rapid descent, the brick-wall lining of the shaft soon disappearing in the total darkness. Jack froze momentarily as the cage hit something, bouncing them from side to side. He gulped again as the tension eased on the rope and the cage swung gently at the shaft bottom.

'Steady it.' Hands pushed hard against the shaft side. Bob eased past the others, and as he made his way down the ladder between the jammed cage and shaft wall, they held it tight. His body momentarily blocked the fiery glow from the shaft bottom, and then it reappeared as he stepped into the underground workings.

'Okay,' Bob shouted. One by one they made their way from the gently swinging cage down the ladder, by the fire-blackened jammed cage

'My God, that's hot!' Bill said. They could feel the heat radiating from the bricks of the shaft wall. All their working lives they had descended down this shaft but nothing could prepare them for the transformation.

'Hell', Eric commented.

Jack's mind pictured the shaft bottom area as it had been, with its high-ceilinged hall and three brick-lined roadways leading away. The floor normally contained a small gauge rail track, and 10 yards down two roads there had been small steam engines. The blast had scoured the brick work around the shaft side clean, and long scratches could be seen. The brick roof still held, but almost filling the roadway was a mass of burning wood, smouldering rubble, iron axles and wheels of coal tubs carried there by the explosion.

'We've gorra make some room,' Jack said as Bob handed his lamp to him. It joined his lamp dangling on his belt; its light was not needed.

'Mind yourself,' said Bob, who was already on his knees, his massive forearms making the shovelling look effortless. The sweat on his bald head was gleaming in the flickering flames. His cap and coat were being held clear by Bill. The sounds of the rubble being thrown to one side brought everyone back to the job in hand.

'Jack, I'ave cleared way to t'water pipe.' Eric almost pushed Jack down to connect the hose pipe. The flames died down as the area filled with hot vapour.

'Take these.' Bill handed his and Bob's coat to Tony.

'Watch where tha's spraying that water,' said Bill in a Staffordshire accent, coughing over his well-muscled shoulder as he joined Bob in the steam.

'First time they've had their backs washed in years,' George said as he helped Eric direct water through the steam onto the burning wood.

'Five minutes.' All stopped and lamps were turned to minimum. Jack's eyes strained to look through the steam at the flame in his lamp, checking for that distinct blue flame which would show the presence of gas. There was no blue flame. 'Clear.'

'That's hot,' Jack told Tony, 'don't burn thy 'tashe,' as they passed large pieces of steaming rubble quickly along the human chain, then up the ladder to the cage and into an iron tub. Once the tub was full, Jack signalled the cage's readiness for the return journey back to the surface with one blow of a hammer on the water pipe in the shaft. The surface acknowledged his signal with a reply of one bang. It coincided with a rumble, which started to rise.

'Five minutes.'

'Clear.' Jack's brain pictured their possible plight if another explosion or a leak of gas occurred – nowhere to find cover and no quick way out until the cage returned. They could all be dead by then. All stood in silence, exhaling into the warm, steam-filled air as the hiss announced its return.

Six shovels were soon working with that steady rhythm that came from years of practice.

'How's it going Jack?' With the return of the cage from its third trip to the surface, Mr Bramley, the engineer, had appeared. Jack could just see the top half of him, the light from his oil lamp creating an aura around him in the steam.

'God has come to Hell,' Eric said quietly, as he climbed on some smouldering rubble to have a better look at us. The apparition then looked all around this hell, and became human. 'Any ideas about the jammed cage?'

'Not 'ad time to think about it,' Jack answered as Mr Bramley scrambled out of view and shouted, 'send some wood supports.' The disappearing of his light, followed by a single bang and a single bang reply on the water pipe in the shaft told Jack that Mr Bramley was returning to the surface.

'Five minutes.'

'Clear.'

'Set supports, George.'

'Why did he have to remind them it was not safe to work under a roof that was not supported?' Jack thought.

'Bring up a couple of 'em wooden props an' a plank cut to size, Eric,' George shouted, flicking back his red hair. Jack held Eric's cut plank to the rock roof just clear of the bricked area, around the shaft side.

'Okay', Eric replied. He positioned a cut post at each end of the plank. They were slightly larger than the distance between the floor and plank; each was then hammered home. The roof was now supported. The thud of the hammer had, for a short time, drowned the scrape and throw of the diggers.

'Put some water on t'shovels,' Bill said.

'You asked for it.' Eric cooled men and shovels – as water hit hot steel, some turned to steam, the rest soaked them.

'Second bath this morning', laughed Adam.

'Quiet,' Jack said at the five minute time and, while he tested, all ears strained for the tell-tale creaking and cracking of wood, their early warning of possible further roof falls.

'Clear, change over!' The sound of shovelled debris hissing, as it hit well-watered rubble, continued, as did their human chain, with the periodic burst of steam as water was sprayed over the area. Then they halted, as the cry came out,

'Jack,' called Adam, who had stopped shovelling and was indicating to a pile of rubble. 'It's clothing.' Jack crawled over. All teams, sensing something had been discovered, stopped work.

'Take it gently,' said Jack as Adam lifted a piece of iron.

'Here.' Adam handed the iron to Bob, uncovering a man's back.

'It's John Trueman, I know those blue zigzag scars on his back. Do you remember Bob? He got them when he got pinned down and covered in a roof fall'

'I dug him out.'

'Sorry Adam, I remember now.' Young Eric was sick as they turned him over. Jack gulped, nearly being sick too.

'Can you all help?' Six pairs of hands slid under his back; six throats swallowed as charred flesh fell away as they lifted. 'Back down.'

'Try that plank.' Bill Burr was thinking of his first-aid training.

'You slip it under him, Bill', Jack said. As we slowly lifted John, more bits of charred flesh and clothes fell away. Turning towards where a coat was draped over the remains of a small coal wagon, he pointed. 'Throw it over, Adam.' They gently wrapped it around the remains of his face and upper body.

'Use this Jack.' Bill handed over some rope. The smell and feel of his partially cooked body made the skin crawl as they tied him to the plank. Jack looked at his team's faces as they eased him by the damaged cage. They would never forget John Trueman.

'Okay Jack,' Bill said as they stood clear. Jack hammered once on the shaft water pipe. All stood as the image of John's body filled their minds. He was now laid on the floor of the spare cage as it ascended, the first stage of his journey to God.

'If we're not careful that could be us,' Jack reminded everyone. For a short while they stood in the steam and listened to the sound of rushing air as the cage made its way to the surface.

'There was no way of warning them,' Jack spoke all their thoughts

'He has a wife and three young kids,' Eric said quietly.

'Yes, I know her, they live opposite side o' railway line from us, in Silkstone Row. I remember them moving in,' Adam said.

'Aye it were a grand day. We moved in same day. There'll be no celebration today,' said George.

'Five minutes. You check, Eric.' Jack handed him his safety lamp and sample-gathering rubber bulb. As he depressed the bulb and took a sample in the most likely place for gas, near the roof, Jack thought, that it would take his mind off it.

'Clear,' said Eric, then he added, 'didn't Mr Garforth invent this special safety lamp? I remember only recently someone saying the Royal Commission on Mine Safety said it were good. Wouldn't they 'ave been using it to test? So what happened, Jack?'

'I don't know lad,' Jack said.

'Okay Jack, shift over,' announced Edmund, the next shift leader. It was an unusually quiet sweaty, dirty team that watched the approach of the evening sky. The stars and clean air announced their arrival at the surface, as they handed lamps and safety tokens in to George.

Mr Garforth passed a warm drink to each of them. 'Okay lads?' There was no answer. He placed his arm around Jack's shoulder and pointed him to his office. 'Tell Harold everything.'

'Is that everything Jack?' Harold asked. For almost thirty minutes, Jack had explained every little detail to Harold; the site and nature of all the rubble, the state of the roof, the position of the body. Somehow he could not think of him as John. Jack drew, along with his fellow officials, a detailed map of what conditions had been like.

'Finished, Jack?' Mr Garforth asked. Jack nodded. Once again Mr Garforth's arm went around Jack's shoulder.

'They've found Henry Deakin'.

'Henry!' Jack cried out. There was then silence as he thought about the fact that they had been friends since childhood. 'Where is he?'

'He's laid out in the store room'.

A few steps from the office brought Jack to the store room. In the gloomy light he saw two bodies laid on the stone floor covered with blankets. 'That one is Henry,' he thought. 'That broken steel segment on the base of his clogs – I have seen its track so many times in the dust of the underground roadway, as he walked in front of me.' After finding John Trueman, Jack imagined what was left of him. Lifting the corner of the blanket confirmed it.

'Henry…' he said, and then stood in silence for a moment remembering their mutual past. Breathing deeply, he said, 'Now I have to go and tell your wife.' As Jack emerged from the store room, dark shapes surrounded him. They had waited. As they walked home together in the clear evening light, their presence eased his sorrow. However, even the clean, fresh air somehow tasted different tonight. One by one, with just a touch on Jack's arm, they left the group to go home. Eventually he reached Henry's and knocked on his door.

'Dorothy, you've heard?' Jack said, as his wife, brushing back her blond hair, answered the door.

'Jack!' She just looked, then hugged him tight. 'Aye I had to come and help as soon as I heard. I'm glad I did – the kids have been crying all night. We've just got them asleep, now Alice is on her own.' Jack heard crying behind Dorothy.

'Alice.' Henry's wife stood there in the dim candle light, desperately trying to hold back the tears.

'Tin bath's ready for you at home, water in' set pot will be warm. By time tha's had thy bath I'll be home.' Dorothy shoved him towards the door. The streets were dark and silent as Jack headed home in the moonlight. Sat in front of a roaring coal fire in the bath he tried to get himself clean. The coal dust would come off but somehow he did not feel clean. The smell of ham cooking brought John's body into his mind.

After the explosion, the long road to recovery began for the families, the village and the colliery. Let me use Jack and Dorothy Dyson to demonstrate that healing.

five

THE COLLIERY RECOVERY

The birds were singing as Jack pulled back the brown bedroom curtains. It had rained during the night and the normally dusty roofs had been washed clean, and the water on the muddy road sparkled, as it reflected the early morning sun. He opened the window – the air was cold but smelt and tasted clean.

'Four more dead bodies have been brought out during the night.' Dorothy stood at the bedroom door. 'However, they say they have not lost all hope for the rest.'

'Who are they?' he asked.

'Jack Gill. Mrs Garforth is helping his wife and their three children.' She paused as they pictured Jack – tall and balding with powerful forearms, he had stoked the furnace at the bottom of the upcast shaft. Its rising heat normally provided the ventilation as cool, clean air was drawn down the downcast shaft, then directed by doors and roadways through the underground workings.

'George Allatt.' She paused again, George was a tobacco chewer; his face had been badly burnt in an explosion when he was young. His spitting would leave a permanent stain where he had sat to operate and feed the steam engine with coal, which pulled the coal tubs through the colliery by means of an endless wire rope.

'He had a wife and two children. She seems to be coping,' Dorothy added. 'William Barker and George Colley.' Their minds were working on similar thoughts as she added, 'I've been told they were both single.' A quick porridge breakfast and a silent hug from Dorothy, and he was on his way through the muddy streets. He passed men and boys chopping wood for the fire and dodged around women hanging out clothes. None called out their normal friendly greetings. Instead, anxious eyes seemed to follow him. The only sounds seemed to be dogs barking, birds, and an occasional whistle of a steam train and the sounds of his boots squelching in the mud. At the colliery the noise hit him – the hiss of steam engines, pulling up and lowering the various cages. Others were driving the machines in the workshops, the long leather belts from their shafts disappeared almost silently into holes in numerous buildings. From these, the thud of hammers or the hiss of saws and other machines could be heard. However, there was no singing or light-hearted banter. He knocked and entered Mr Garforth's office. Mr Fisher's unshaven face turned towards him.

'Morning Jack. Cup of tea for us both please Harold, while I update Jack.'

'Morning Mr Fisher,' he answered and thought, 'he's not been home.'

Mr Garforth was tracing with his finger an area of underground roadway on the diagram on the table. A strange gentleman standing next to Mr Garforth gave a short cough.

'Jack let me introduce you to Mr Wardell, the mines inspector.' Mr Fisher pointed with his arm. Mr Wardell, a tall, thin gentleman with a blue scar just above his left eye, took a pace towards him. His grip and the look in his eyes as they shook hands told Jack that this man cared.

'Can I emphasise the importance once again of what Mr Garforth said yesterday of making careful notes of everything?' Mr Wardell said

'Yes sir,' he replied.

Mr Fisher nodded towards the door and as they walked outside. He explained, 'Mr Wardell arrived on horseback at four o' clock this morning. He and Mr Garforth have already been underground twice.' Then he added, 'I suppose you've heard about Jack Gill, George Allatt, William Barker and George Colley?'

'Yes.'

'You know you're not on shift again till this evening.'

'Aye, but can I help in any way?'

'The best way to help, Jack, is to be alert when you go down with your team, so go back home and try to relax.' He'd just started walking away. 'Jack,' called Mr Fisher, 'have you heard about Ned Kaye?'

'Aye, 'im and his lad Allen got out okay then he collapsed and died. Allen took him home in a sack.'

'Well, Allen and Sarah, his mother, stripped Ned then laid him out ont' table in front room in his best suit, shirt and tie,' Mr Fisher explained. 'Early this morning Sarah heard a banging down stairs, she woke Allen up and when they opened the front room door, they were amazed to see Ned sat next to table with his tie and one boot off. When he saw them he said,

'I'm not dead yet lass.'

'Anymore news?' He was asked that question many times as he walked to his allotment garden. They all knew about the additional deaths so he repeated the story of Ned Kaye – it brought a wry smile to their faces. After a good two hours of digging and turning over the heavy ground, it felt so good to be alive. He walked over to talk to men who had had a similar burst of energy, but were now grouped together.

'Have you heard about Ned Kaye, Jack?' Isac Bednall asked.

'Yes I've been to pit. Mr Fisher told me.'

'Any news on how long pit's going to be shut? Any jobs going at other pits?' James Halam asked.

'All the bosses are too busy to even think about that now,' he answered.

'What do you think, Jack? My Sarah says when her dad got killed in an explosion, they had to go to Newcastle and beg for food on streets,' said Jacob Doody.

'Nobody from village will have to go to Leeds or Wakefield and beg – we'll look after our own.' Nodding heads agreed with Jack's words. On returning home Dorothy stopped her ironing, and placed the flat iron on the rail in front of the fire to keep warm. Jack sat down at the well-scrubbed kitchen table. Without a word she took a large pan off the same rail and poured stew into two dishes, one of which she placed in front of him. Then replacing the pan she sat at the table opposite him.

'Alice is taking it hard, she can't stop crying.' He looked up as she spoke. 'Her eldest daughter Freda is looking after them.'

'Freda! She's only ten!'

'She's grown up quickly. She fed them all, then sent the other five kids to their aunt's with a few clothes.' She blew on a spoonful of stew.

'They live in Normanton.'

'Aye, the kids will miss their home, but the aunt will be happy to look after them for a couple of days to give Alice time,' she said as she dipped a thick slice of newly baked bread into her bowl. They talked about the children. The two youngest of Henry's could not understand what had happened but the other four had withdrawn into themselves and only answered when spoken to. 'Aren't you going to bed?' she asked eventually.

'I'm not sure I'll sleep,' he answered.

'Jack, you can't afford to be tired on your shift. I don't want anything to happen to you.' She shoved him towards the stairs, then sat quietly sewing and repairing his clothes as he had a couple of hours sleep.

In the cold of the early evening men conversed quietly as they made their way to the colliery. Very few candles or oil lamps gave light from homes – people were keeping costs down. It was Mr Garforth who updated them on the situation. He came to the door when he heard the sounds of clogs stop outside his office. 'Have you heard they're just bringing out young Tom Ibbetson?' Jack didn't know about Tom.

The briefing over, they walked to the shaft side. There, in the dim light, Jack almost stepped on the headless, armless trunk of a body with only one leg beneath the remains of short trousers. He swallowed hard, and then, trying to shield the body, reached down and picked up a sack from a pile nearby, and draped it over the corpse.

'Tom Ibbetson?' asked Bill Burr as he stepped over it. No answer was needed

'Sorry lads,' Mr Fisher emerged from the shaft-side office. 'We were waiting to see if they found any other parts of him before we sealed him in a coffin.' They stood in a circle around him.

Tom had whiled away his hours of work singing hymns, as he sat alone without even an oil lamp in the total darkness, listening for the sound of the ponies approaching his ventilation control door.

'Can you sing "The Old Rugged Cross"?' Jack had asked one day, when for a few seconds he had transformed Tom's dark silent world as he passed by.

'Sure Jack,' said Tom 'why?'

'It would have been my father's birthday today and it was his favourite hymn.'

'How did he die?' Tom asked.

'Long hours working in dusty and poor air,' Jack answered, thinking of how his father's once powerful body had wasted away slowly, until he had decided that no longer would he be a burden on his family, and the fire had gone out of his eyes.

'On an old rugged cross … ' Tom's soprano voice rang out along the dark narrow roadways.

'What will God and my father make of that?' Jack had said, leaving Tom to his lonely vigil. Some oncoming ponies had slowed and pricked up their ears, as his soprano voice trembled on the high notes.

'We'll miss you,' George said, and they bent their heads in silent prayer. The cage signal bell ringing brought an end to the moment of respect. They turned and made their way towards the rattle as the screen on the front of the cage was lifted. The whistling air as they descended shut out the last notes in Jack's mind. The swallowing allowed their ears to pop with the change in air pressure and brought them back to reality. Within minutes of climbing past the jammed cage, Jack's team had established its shovelling rhythm. A smell slowly settled around them.

'What's that smell? I must get some fresh air.' Bill Burr stood up, and then as he realised it was the rotting flesh of the ponies, he doubled over and was sick. Eric and Tony joined him as they all gulped, then ran to breathe in some clean air just past the road that went into the stables.

'I'll see boss when we get out – we can't work in that smell,' Jack told them, as they returned to their work after a second movement to cleaner-smelling air. On completion of the shift they had cleared 50 yards. The teams clearing other roadways that branched out from the shaft bottom had cleared about the same.

'Mr Garforth and that inspector fellow went home last night for the first time since the explosion,' Mrs Arnold told them as they passed her doorstep on Monday. 'More bodies had been found. I don't know who.' On arrival at the pit they crowded into the cold storeroom where the bodies were laid.

'Who are they?' Jack asked a dark figure bending over new bodies.

'George Cookson and Charles Plimmer,' Mr Buxton replied. He thought, 'George is single and Plimmer I don't know.'

'Bartholomew,' Mr Buxton called his fellow official as he moved outside, 'can you and your team start bringing out the dead horses?' Jack inhaled, deeply remembering the smell. One or two of Bart's team turned a sickly white.

'Thank goodness it's not my team,' he thought.

'Your turn tomorrow, Jack.' Mr Buxton spoiled his congratulations, then continued 'It's double pay while you're doing it.'

The crowd was now smaller and made up mostly of strangers. They parted to let them pass through to Mr Garforth's office. As Mr Fisher watched them approach he said, 'They are the only ones we have discovered, so there is still hope the others have been trapped in a pocket of good air.'

As Jack looked towards Mr Wardell, Mr Fisher added quietly, 'It was dark when the inspector rode home; he can't have had much sleep. It's nearly forty miles, and there's no trains. He was back early this morning on a fresh horse.'

They crammed up tight to each other in the cage; the smell of decaying horses rose and grew stronger as they neared the bottom. In the light of the acetylene torches, Jack noticed the wet cloths moving in and out over mouths and noses, as men cut up the jammed cage.

'Well it looks more like pit,' George said. All the rubble had been cleared away, a small steam engine was pulling tubs full of rubble to the shaft side. As they walked carrying two water bottles the 200 yards of cleared roadways, their wet cloths seemed to cut down the smell from the stables.

'Leave some for us!' Jack's joke was not appreciated by Bartholomew as they branched off towards the stables. He asked, 'What's the straps for?' pointing at large leather straps they were carrying.

'Yesterday the horses fell apart when they lifted 'em. It weren't pleasant. That's what straps and these are fo'.' Bartholomew held up large leather gloves before disappearing into the stables. Once past the stable road, the incoming air was almost breathable.

'With all that hay in the stables it took twelve hours to put fire out,' said Tony, 'and they had to work in that smell all the time.'

In spite of the damp cloth causing breathing difficulties, the team soon settled into their routine. Jack began thinking about George Allat. George had been a pillar of the local church and conductor of the 100-strong choir. His wife had been given the bible he always carried; its charred remains had helped identify him. He would have led the singing with his loud baritone voice last night. Instead it had been a short, solemn service. The thud as Matthew's body hit the floor brought Jack's mind back to reality.

'Gas!' he shouted back towards the shaft side. He staggered a few steps towards it before joining the rest on the floor. Adam had been returning with a shovel which had been left in the short connecting corridor that joined the two shafts, when he heard the cry.

'I've got to get help,' he said to himself and returned through the doors to the downcast shaft side.

'Nathaniel,' Adam shivered in the cold incoming air, 'gas, they've all been overcome.'

'Right Jas, you inform the surface what's happened, t' rest come with me.' While James Whittingham signalled for the cage, Adam, Nathaniel and his team moved into the connecting tunnel.

'Just a minute, I'll test for gas.' All stood and waited whilst Nathaniel carried out his test. 'Clear!' Eight men took a deep breath.

'Adam and I are going into the return air now. We will get a body and bring it back here. No-one else is to go through those air doors until we come back − is that understood?' His team nodded as Nathaniel looked at each one in turn. 'Adam ready?' Adam nodded. 'Okay, deep breath. When I say right, Moses you open the air door and shut it the second we are through, then listen for us returning.' Moses Perry moved over and held the wooden handle on the heavy air door.

'Right.' Both took a deep breath and ran through the open door. With mouths firmly shut and holding their breath they quickly reached the first of the bodies − it was Jack. They each grabbed one of his lifeless arms, pulled it over their shoulder and, with him suspended in between them, ran for the safety of the connecting tunnel.

'Moses, take over from me.' Nathaniel moved back to the door as Moses moved in, placing Jack's arm over his shoulder. 'James and John ready?' Both took deep breaths. He opened the air door. 'Go! Moses, get Jack into clean air at downcast shaft,' he said as he shut the door. One by one the overcome men were removed into the good air, and, as the cage arrived, they were transported up to the surface. The waiting crowd had seen the men running to the office and their quick return. Rumours of a new disaster started as the first of the bodies was carried out of the cage. They quickly passed through the village.

'Who is still missing?' asked Jack when he recovered.

'John Newton,' came the answer.

'Wife and five kids,' interrupted George Hopwood with bitterness in his voice. 'Thomas Oakley, he's single. James Nicholson, wife and two kids. Martin Buxton and John Fox.'

'They got married same day as me,' said Eric.

'Abraham Davies.' The voice paused.

'William Bradall and his wife Elizabeth from Silkstone Row co-op shop are helping Abraham's wife and their six kids,' Adam Firth added. 'William and Abraham have been friends for years.'

'Fifteen-year-old Sam Flint, young George Wilcock. George Megson and Solomon Green, both married, I'm not sure if they've any kids. And Samuel Lomas,' the voice completed reading the list.

'Didn't Sam do his official's training with you, Jack?' asked Bill Burr.

'Yes, I know his wife Ruth and their four kids well,' he replied. 'There's still a chance they're alive.' No-one spoke. They all knew there was no chance.

When Jack arrived at the colliery on Tuesday evening, the rain and cold wind had driven away almost all the spectators.

'We are going to concentrate on getting the horses out. I need a volunteer official and twenty men.' Mr Fisher said as soon as all the teams had arrived.

'Okay,' Jack said. 'it was my turn anyway. I'll organise volunteers.'

'Thanks Jack,' Mr Fisher said.

'Any volunteers?' Jack turned and addressed his team. All just stood and looked at him, no-one answered, and then he smelled the disinfectant and looked down. They were already carrying the large leather straps and gloves.

'Nice day for it,' Bartholomew remarked as they gave their boarding tokens with tied wet cloths over their faces, and entered the cage, 'you'll be glad of the smell of that disinfectant. We've sprayed it everywhere.' Crammed in the cage the smell was quite overpowering, and it was worse at the shaft bottom. As Jack entered the stables, he crunched the burnt skeleton of a mouse and more small skeletons crumbled under his clogs as he moved into a stall.

'Right lads – the sooner this job's over the better. Eric and George, bring that tripod over here.' Jack turned to Bill Burr, 'Bring that lifting block and tackle Bill.'

Jack helped Bill lift the block and tackle on to a hook beneath the tripod. He and Bill then forced their hands through the flies and under the decomposing horse, linked hands and lifted; George and Eric did likewise. Tony and Bill Harrison quickly fed the leather straps through, then hitched them to the bottom pulley of the block and tackle. They all stood clear as Bill Burr started pulling on the lifting chain. Now they really disturbed the flies and the air was black as they buzzed about them. They were relieved that their noses and mouths were covered. 'Bring waggon,' Jack shouted through clenched teeth, as the remains of the horse now hung with its hooves touching the ground. He and George held the legs to one side, as Eric pushed the large flat waggon in place. Slowly the horse was now lowered by Tony and Bill Harrison, as the rest strained on another rope attached to one of the horse-holding belts, so that they could pull it over and lower it down.

'One down, forty more to come,' Adam said, ignoring the flies as the waggon was pushed to the shaft side.

'Thank goodness cage is now okay,' Eric said, thankful that they didn't have to start the lifting process again to get them past the jammed cage. His sentiments were echoed as they pushed the cooked rotten flesh, swarming with flies, straight into the cage.

'What'll happen to 'hem?' Eric asked as the cage started its return to the surface.

'It might shift some of them gawking strangers when they're carried out', George said, as the flies buzzed past to return to the stables.

'There's a large pit been dug for them at side o't railway. Now let's stop yatting and get rest out,' Jack said, asserting his authority. At the surface William the banksman, hardly recognisable with the wet cloth over his face, pushed the flat waggon and its occupant off the cage and towards the crowd, who started a rapid backward movement. He was soon joined by other men with similar face masks, and the body was moved to its last resting place.

Meanwhile, Bartholomew's gang had set off over the partially cleared roof fall, looking for survivors. Jack was told later that they had walked or crawled in single file as quietly as possible. Bartholomew was a short distance in front calling out, 'Anybody there?' Then a loud crack sounded at their side as a partially burned wooden support gave way. Bartholomew looked around, a second crack followed as the tunnel, now unsupported, began collapsing under the enormous pressure. Wooden roof supports on the other side of the tunnel were pulled from the perpendicular to an angle of 60° and then 45° and lower as the roof dipped sharply. Bartholomew scrabbled forward as roof and floor slowly came together. His team ran backwards and, with a final whoosh of trapped air, rubble and dust surrounded them as roof and floor met. When the dust cleared, very little air was forcing its way past the rock fall to Bartholomew. He was trapped.

'I'm okay!' Bartholomew shouted through the rubble.

'Right, we will soon have you out,' called Benjamin Day, as well-muscled shoulders lifted and threw the heavy rocks to one side. Others ran back to the downcast shaft bottom to get wooden props to re-support the roof.

'Hold it lads!' Bartholomew shouted through the almost solid wall of rock. 'I've just been checking with my lamp. We've got gas coming down – leave me and go.' Arms worked faster – the roof fall had almost destroyed the natural ventilation; time and oxygen were precious. Streams of sweat ran down chests that were breathing deeply. Torn fingers wiped moisture from eyes as they set new supports. Eventually the ventilation started to increase as holes appeared in the blocked road and they could get through.

'Is he okay?' John Hopwood asked the rest when they had eventually gathered around the unconscious Bartholomew.

'No lad,' James Cranswick told him, 'grab a foot. We've got to get 'im to fresh air as soon as we can.' In teams of three, changing when one tired, they carried or dragged Bartholomew towards the shaft bottom, the air getting cleaner with every step. Meanwhile, Jack had called a halt, and he and his team had walked through the connecting tunnel doors to get some clean air in their lungs from the downcast shaft. Seeing Bartholomew's team coming, he ran towards them. He circled around behind Bart, placed his arms around his chest, squeezed and let go, then again and again. Bart coughed and his chest began to move.

'You've brought him back to life Jack!' John Hopwood said. He seemed to be the only one who could speak; the rest just stood there breathing deeply.

'Help me sit him down,' Jack said to George. Bill Burr, meanwhile, had signalled for the cage, and on its arrival he looked at Bartholomew's exhausted, sweaty team.

'Can you manage to get 'im out?' Bill asked.

'Sure,' answered John. They surfaced an hour later. 'How's Bart, George?' Jack asked the one-armed banksman.

'He's fine but if you don't get your smelly bodies out of here we'll all be sick,' George answered, signalling to the waiting team to enter the cage. They went over to pay their respects to the ponies. The thin coating of lime covering them kept the flies down, but the crows were waiting. They had known each one by name and personality, and all had their own favourites who they had taken tit bits. They would be sorely missed.

'Jack, can you bring your team along with Bartholomew's and Nathaniel's into the office?' Mr Garforth called as they arrived next day. Once they had all crammed in, he continued. 'We have now made the decision that anyone left underground is dead.' His slow speech and the look on his face showed how he and they all felt. All stood in silence for a moment. 'No more heroics. We don't wish to lose anymore lives, okay Bartholomew?'

'Yes, Mr Garforth.'

'Right Jack, an electrical ventilation fan has arrived from St Johns Colliery at Normanton, along with a group of experienced men. It's at the upcast shaft bottom slung under the cage. You and your gang help get it in position.' He turned. 'Nathaniel, you're on double money today. Bartholomew, carry on cleaning up.'

'Well we certainly have variety in our work,' said Adam as they left the office. 'Now we have foreigners telling us what to do.'

'They know how to set it up,' Jack said. 'You know we've never had need fo' that type o' ventilation. Furnace at bottom o' shaft has supplied all ours.'

'Bartholomew, get your lads to load timber and tackle over there ready to go down pit,' said Mr Buxton.

'It's marked Briggs Collieries,' said John Hopwood as they arranged to lift the bound timber from a horse and cart on to a flat waggon.

'Aye, all the local collieries are giving us all the help they can. All the managers are in constant touch. Most were here within hours and offered help straight away,' Mr Buxton pointed out.

'Jack,' called Mr Garforth, who quickly moved over to him. 'Let me introduce you. This is Mr Marshall Nicholson. The engineer from St John's Colliery can't come; Mr Nicholson is the engineer from Middleton Colliery near Leeds.' They shook hands.

'Right Jack, do you know how fans work?' Mr Nicholson asked.

'No.'

'Once the blades start turning they push clean air forward,' Mr Nicholson said as they moved out of the office. 'We direct the air to where we require it by using those large pipes.' He pointed towards wooden pipes. 'Can you make sure they all get down without getting damaged? I will join you down there in a short while.' He disappeared back into the office.

The shift was spent getting the wood in the cage, and then the large wooden pipes, one at a time, were slung under the cage that had been raised right into the headgear. At the bottom, Jack's team waited, and as the pipe swung gently they got hold of it. By controlling the descent of the cage it was eased out of the shaft and laid in the roadway.

'Now Jack we have to mount the fan and pipes on those wooden beams,' said Mr Nicholson, who had travelled down with the last of the pipes and indicated some cross-like wooden structures. He measured out the distances for them to place the fan and the structures, and then they pushed the interconnecting pipes and the fan together. Meanwhile, Mr Nicholson had been connecting a cable to electricity terminals at the bottom of the shaft. 'Everybody stand clear!' Mr Nicholson shouted. Everyone moved and watched as he pulled down two handles. The blast of air as it was turned on caught them by surprise.

'Now we can start to recover the colliery,' said Mr Garforth, who had joined them.

'Only twenty miles of underground roadway to do,' Jack added quietly. Slowly they advanced underground, removing the gas, rubble and bodies. The bodies were lifted using bratish cloth and placed in special boxes that had been made with carrying handles. Once on the surface they were placed in polished oak coffins made extra large to accommodate the boxes. Teams of men transported water containers to douse the fires.

On 18 October the smell of burning became very pronounced as they advanced down one roadway, and Mr Garforth was sent for. Along with Messrs Fisher and Buxton, he was soon on the scene.

'Jack, your team is doing a good job there,' commented Mr Garforth. 'Can you leave them and come with us?'

They travelled towards the smell, and periodically Mr Garforth, who was in front, banged his dust-laden trousers to ensure they were going with the ventilation instead of into it. They then began to feel the heat.

'Right, 10 yards between us. Jack, you be back marker,' said Mr Garforth. Then he walked the few yards to a roof fall, and dropped down. Slowly, he crawled forward over the fall, repeatedly saying 'I'm all right.' In the rear, as Jack crawled forward, he periodically stopped, listened and looked. A cracking sound or small pieces falling from the roof could herald a rock fall. Eventually looking forward he noticed Mr Garforth's lamp had disappeared.

'Wait!' The message was passed to Jack by Mr Buxton. Lying in the small space between the floor and roof left by the fall, Jack waited. His safety lamp held out in front of him, listening and looking and every few minutes taking samples of air with his bulb, watching for that tell-tale blue flame, as the sample was gently squeezed over the flame. He tried to raise his head to see what Mr Garforth was doing, and banged it on a sharp piece of rock. 'Right Jack, we're going back,' Mr Buxton said, as he turned around and crawled towards him. On the way out of the colliery, over the noise of their clogs, Jack heard Mr Garforth say,

'We are going to seal off each area of the colliery in turn and stop the fires getting oxygen. This area first. We'll look at the plans for the best place for the stopping when we get out.'

'It's a good job there's nobody left alive,' Jack thought to himself.

'By the way,' Mr Garforth spoke to Mr Fisher and Mr Buxton so they all could hear, 'never again will all three of us go underground together. If an accident happens, all the top management will be killed.'

Now the village, with help of Mr Edward Cowey from the Yorkshire Miners Association, started a collection for the bereaved families. Mr Garforth and all the colliery officials donated large sums. Most of the miners had been on low wages for months. Sam Flint had only been taking home six shillings a week.

Many families now faced months without any income, so Mary Garforth organised the village.

six

THE VILLAGE RECOVERS

The train came screeching to a halt as if announcing its arrival. The thick cloud of smoke from its boiler mingled with smoke from other steam trains, curling around the sooty ornamental iron works that supported the glass roof which protected the passengers as they alighted at Normanton station. Train doors bounced open as station personnel in many different company liveries shouted information and people talking excitedly streamed to and from the passenger compartments. Then a porter, dressed in his red coat and hat, trimmed with gold lace, shouted,

'Parcels for Pope & Pearsons relief committee.'

'Thank you porter,' Mary Garforth answered, as she and her group of ladies pushed their way across the platform towards him.

'Sign here please.'

She signed while the others collected the parcels, and then they made their way off the smoky, crowded platform, up the station stairs and across the bridge. Darkie, her carriage horse, was disturbed by the constant blowing of train whistles and the steam and a white faced young man held him.

'Me mam said I had to hold him and calm him down, he's all right now,' he said then without giving his name he wiped his nose on his dirty sleeve and disappeared into the steam.

Mary's cellar was now full again with parcels. There was a knock at the door and her maid entered, 'Would you like anything my lady?'

Normanton Station. For many years this station was amongst the busiest in the country.

Wash Day. For the miners' wives, with husbands, sons and lodgers working at the local collieries, every day was wash day.

'A warm drink would be nice, and when the ladies start arriving please let me know.'

'Right ladies are we all ready?' Mrs Garforth's question halted the many individual conversations. 'Can we stand for a minute and pray for the six bodies that have been brought out today.'

After a short silence she smoothed out her dark brown dress and sat down. Mary Garforth waited while the ladies also sat, then said, 'Can we go around the table and see who has anything to report. Caroline, you first.'

Caroline Wilmot brushed her brown hair away from her forehead, looked around the large, well-scrubbed table in Mrs Garforth's kitchen and said,

'We distributed the goods that came from our friends in Manchester. We have some more coming this Thursday that Mr Pickard has been collecting in London.'

'What time will they arrive at Normanton station?' asked Mrs Garforth.

'6.30 in the evening, Mrs Garforth,' she said after checking her pencilled notes.

'Can I ask you once again, please call me Mary,' and she looked around the table. 'Who can help?' Several ladies raised their hands. 'Betsy, have you anything for the meeting?' Mrs Garforth said after jotting down the names.

'Agatha Thompson needs help,' Betsy Hughes replied as her well made body shook.

'What's wrong with her?' asked Mary.

'She's just tired out, we've managed to coax her into bed,' and Betsy demonstrated how she had carried her.

'I'll do her washing,' volunteered Dorothy Dyson, thinking that Jack wouldn't mind.

'I'll take the kids during the day,' said Eliza Johnson. Her house always seemed full of children.

'Me and my daughter Sarah will clean and look after Agatha,' offered Elizabeth Newton. It was an opportunity to repay her for past help.

The sound of Mrs Garforth's pencil could be heard as she made notes.

'Alice Lomas is still not speaking. She just sits there, we can't get her to do anything,' Hannah Whittingham informed the meeting in her high pitched voice.

'How long is it since we buried her husband?' asked Mary.

'Six weeks,' replied Hannah.

'I'll pay her another visit, and see if I can do anything,' said Mary.

As the ladies reported the problems of the village, they were discussed and solutions organized prior to returning home.

'Jack, Jack, help me' came the cry. He moved rapidly to the door and pulled it open, Dorothy, his wife, stood there holding a pile of blankets with the wind blowing her skirt. 'Well take them before I drop them!' He moved across and took the pile from her and put it onto the kitchen table. She turned and disappeared and Jack returned to his seat in front of the roaring fire.

'Is the water warm?' she said, dumping a pile of clothes on the stone flagged floor

'Yes, I've filled setpot up and taken the ashes out,' Jack said over his shoulder.

'Agatha is ill,' she said as she ladled hot water from the setpot and poured it into the washing tub.

Jack waited for the sound of clothes and carbolic soap rubbed up and down the scrubbing board to stop and asked,

'I thought you were at a meeting with Mrs Garforth?'

'I was, Jack,' Dorothy replied and, arms deep in the soapy water, she thought for a while, 'Mrs Garforth asked us to call her Mary but she's the boss's wife, we can't do that can we?'

Jack's wife was involved with the other rescue. 'She says her and Mr Garforth will never forget seeing starving children shivering in freezing rooms.' The crank of the roller stopped to be replaced by the squeak of pulleys as a wooden frame was slowly lowered. 'I'm beginning to understand what Mrs Garforth means. It was terrible them first visits to families whose men were dead or missing. We just sat and held their hands.' Tears came into her eyes as wet clothes filled up the frame, 'It's worse now Jack, they are all reacting in different ways. Some just cry all the time, some act as if nothing has happened, others have taken to drink. I feel so helpless.' The squeak returned as the frame was pulled, full of cloths, towards the whitewashed roof and tied securely.

'The kids run to us when we go, they snatch any food we've got, and hide in corners to eat it. Some of the wives have just given up. Their kids are dirty, babies cry for milk their mothers can't give because they've not had a good meal for months.' He moved to her and she turned and wrapped her soapy arms around him. They stood in silence until the fire spitting signalled her return to the wash tub. He knew he wasn't going to get any sleep, so he up picked up the empty coal bucket and said, 'Can I help?'

Each miner washed off the dust in front of the fire. Heating the water was no problem; coal was supplied free from the colliery.

This is what forged the community together. The ones with, helped the ones without. It was happening in countless mining communities throughout the country.

A room at the colliery school was used to store food. The children were a bit crammed doing their lessons, but they managed. One old gentleman said someone had offered to leave a couple of ferrets in the food store to keep rats away, but was told it would distract the kids from their lessons.

The men of the village met weekly. They stored all the produce from their allotments so it could be shared out and they scoured the countryside for wood to chop up and distribute. They discussed how James Worthington had died, never recovering from his burns. The kids were not forgotten, and games were organized for them at the weekends. The men also decided that all the coal delivered free by the colliery to families who had lost men, injured or dead, was not be left on the street.

A rat dropping dead after drinking out of a local stream stirred Mrs Garforth into action and she arranged for the foot deep sewerage to be cleaned away. Families who could not afford to buy from the water waggon that came round had been using the stream water. Twenty-six died as smallpox ravaged the village.

Money was raised through numerous collections at mines nationwide and Mr Pickard and the Miners Association sent food parcels donated by London families, who had read about the disaster in the newspapers. There was widespread sympathy, as people were aware that it could be their mine to suffer next. The carbolic acid and sulphate iron used to clean the colliery was also used to clean the village.

'Fairs come to Normanton,' Dorothy told Jack, 'and we're going to stand outside.' The previous year the annual fair held on a Saturday night in Normanton Town centre had been crowded with drunken men and women. With the heavy rain the mud had been so deep that in the dark, one unfortunate had lain there unnoticed, covered in mud, and died.

The fire was roaring up the chimney as Jack finished a bowl of warm soup, and settled in his rocking chair for the evening. As the cold wind rattled the door, Dorothy pulled on a pair of old boots and asked, 'Are you ready?'

'I'll get me coat.' Jack moved towards the door.

They met up with other helpers and Mrs Garforth as they walked along the frozen, mud road towards Normanton, some holding oil lamps. In the dark, a train passed under them on its way to Normanton Station. Emerging from the steam, they looked at the gas-light station on their right.

'Look, elephants are being moved from t' stables at the Midland', Dorothy pointed out as they approached the town centre. Noise and light seemed to hover over the two public houses, the Midland and the Junction, which faced each other across the busy Market Street.

'Quick, let's get to the circus before the crowd,' Mrs Garforth urged.

Between the public houses a large crowd, almost knee deep in mud, parted to let the elephants pass through from the stable they had been sharing in the Midland pub yard. 'Dad, what do they eat? Jimmy told me they'd eat me,' a young lad asked, pulling his dad's hair as he watched from his perch on his father's shoulders, the flicker of the gas street lights reflecting on his excited face.

'Dad, why do they have two tails, and why do they hold each others tails?' asked a young girl from a similar perch. A fanfare of drums drowned the reply. Burning torches were held high and heralded their arrival as mud splattered while they ambled into a large circus tent, pitched on the grass behind the Midland.

'Jack, you load supplies into the trap.' Dorothy said as they waited in the light of the torches. He was soon busy loading chickens, sides of ham, bread, baskets of apples, pears, plus other fruit which appeared, filling up the trap.

'We've collected 7s 9d,' said Mary Webster. Even money had found its way into a shaken bottle.

Two lads staggered through the mud with a barrel of beer, and heaved it into the driver's seat, 'Landlady from the Junction sent it,' one grunted. Mrs Garforth held the reins in her hands as they walked back in the dark to Altofts.

Planning had already started for what they were going to arrange for the kids for Christmas. The women were already knitting dolls, whilst children had given up their toys made the previous year. Men were spending any spare time carving dolls or small boats.

On Christmas Eve, Mr Garforth lent them a pit waggon to distribute the toys. On New Years Day, the local police constables dressed up as convicts, clanking ball and chains attached to their ankles and shaking bottles to raise money, touring the town. This act was just one of the many different ideas which were explored at meetings at Mrs Garforth's Halesfield home, either for raising money or just for cheering up the village.

The work of recovering the colliery was completed by 3 December; it took many years for the local people to recover.

seven

1887-1890
THE CONCLUSIONS
ARE RIDICULED

White ice stuck to Jack's clogs as he crunched through the soot-blackened snow in the early morning light to the colliery lamp room.

'Mr Garforth wants to see you, Jack,' a voice spoke, and with a rasp, the dim light was illuminated by an induced spark that caused an oil lamp to glow. He turned and saw the dim glow of gaslights in Mr Garforth's office, even at this early hour. He crossed the now-busy colliery yard, stamped his clogs and hit his cap against the wall, then knocked.

'Good to see you Jack,' Harold, Mr Garforth's clerk, said, then he immediately turned and gave a short rap on Mr Garforth's door before disappearing into his office. Jack was brushing away the last of the snow with his hands when Harold waved him in.

'Jack,' Mr Garforth said, looking up, 'thank you for coming, sit down.' Harold brought a brown leather padded chair for him. 'Two pots of tea, Harold.'

'Me, sat drinking tea in boss's office,' Jack thought as he sipped his tea.

'I need an experienced mining man who is practical, honest and hardworking.' Mr Garforth stopped speaking, and looked at Jack. Then filled his pen from an ink bottle, signed his name to the document he had been writing, and stretched his neck above his stiff white collar and black tie and continued, 'He will work on various projects I have in mind, to improve the safety of the mine. Will you help?'

Jack's pot of tea stopped half way to his mouth, and travelling down again, nearly missed the table. It was only the thud, and his reaction to the pot tilting, that helped Jack's astonished brain to focus.

'Yes.' How could he refuse?

'Good. Have you heard there's been a similar explosion at Elemore Pit, twenty-eight dead?'

'Where's that?' Jack asked.

'Near Netton, in County Durham. So, no time like the present to start. These are all the reports concerning the explosion in the Silkstone Pit,' Mr Garforth stood up, brushed a few specks of dust from his black jacket, then collected a bundle of papers from a walnut wooden cupboard. He stared at them for a long while, and then continued, 'These are the notes made after the explosion. Arrange them, study them, and consider the roadways prior to, and after, the explosion.' The papers thumped onto the table in front of Jack. 'Write out your thoughts, no matter how silly they sound. I think I have an idea what may have caused the explosion.' Jack pushed the papers into a neat pile as Mr Garforth opened another door in the cupboard and pulled out a wooden box. 'I took these after the explosion. They may

Underground roadways at the West Riding Colliery.

help,' he said. A box of plate-glass photographs rattled as he put them on the table and Jack placed the box on the papers. 'Harold will show you a small office we have prepared. Harold!'

When Harold entered, Mr Garforth continued, 'Show Jack to his office.'

After reading through the notes the following morning, Jack went underground. Almost every yard as he walked had a memory for him. As a young boy, he had nursed his candle, seated at his post by that air-door; it controlled the ventilation for this roadway. The slap as the pressure of air forced the ventilation door shut behind him reminded him of the times he had sat in total darkness, hearing every creak and groan from the roof-support timbers, listening for the sound of approaching ponies. Their clattering hooves on iron rails betrayed their nearness. Quickly, he had opened the massive wooden door and held it against the pressure of the incoming air. Then a ghost-like vision of a pony, man and tub, illuminated by the pony driver's oil lamp, would glide by. He could hear its deep breaths and feel the warmth of its sweaty body as the dark swallowed the diminishing clatter of hooves and tub wheels. Occasionally the pony drivers lingered for a chat and to light a candle, but mostly the ponies plunged on with the full tubs towed behind them. In the darkness, he could not even make his way out of the pit until someone came to collect him.

Jack continued his walk up the haulage road, stopped, looked carefully around and, folding his legs beneath him, sat down. As younger lads had started work, he had progressed to working this haulage road. Here, in the light of a hanging oil lamp, he quickly learned to transfer coal tubs from ponies, or one continually-moving wire-rope haulage system to another. Speed was essential as they unlashed the small chains that hooked the tubs to one rope and then lashed them to the new rope, always aware of the danger as the chains tightened. He had lost the tip of one finger; many of his friends had lost fingers. One had been half asleep and the ever-moving rope rattling over rollers had clamped tight, the lashing chain on his arm; he would never forget his cry.

He pressed down with his hand as he started to rise, then stopped and looked down. No dust! He distinctly remembered the sound as his clogs had disturbed the dust between

It was noted that wet areas stopped the explosion from travelling.

the sleepers, holding the iron rails as he passed previously. Dust had been everywhere, on the thick wooden props that supported the roof, and the planks that lay between them. Some supports had cracked under the enormous weight above them. The resulting crooked splinters of coal and wood standing out at dangerous angles for the unwary passer-by had also been covered in coal dust.

Ears strained as a crack and a few dropping rocks announced that the strata above was now settling onto its new wooden supports, the old ones having burnt away in the intense heat of the explosion. He noticed that there was no dust between the supports either. No dust was raised as he moved on.

It was just off this corner he had hewed his first coal. How proud he had been, to move onto the coalface and help his brother. He remembered the echoing sound of their picks and shovels as they lay in the 12in-high space they had cleared under the coal. In the light of their candles, he had pushed in the small supports that enabled them to go in over a yard under the coal. Later, these same supports had been pulled out, allowing the approximately 1 yard-thick seam of coal to drop. The rock roof scraped their backs as they bent under the new roof to set wooden supports about 4 feet high. They carried or shovelled the resulting lumps of coal towards the enlarged space created for the tubs, and there they had gently put the coal lumps into the waiting tubs.

'It's only big lumps we get paid for' his brother had told him many times. Once, full and with the family token on the tubs, they had been pushed to where they could be harnessed to a pony. If a pony was waiting, it was given any titbits the men had left from their food. Each one had its own character, and the men came to love and respect them. The waiting empty tub was collected on the way back.

It was hard, dangerous work and the scars on Jacks body testified to that. He had been lucky and gained the respect of the manager. Now he was in a position of responsibility.

Jack continued his tour of the underground roadways. Periodically he stopped, hung his oil lamp up and in its glow, made more drawings and notes. Once back in the warmth and safety of his office, the information was carefully drawn on to large sheets of paper, then he added the position of the bodies.

Over the next few weeks he visited all the underground roadways where the explosion had travelled and when he had finished, informed Mr Garforth.

'Right let's have a look at what happened,' Mr Garforth said, moving over to the table as Jack spread the first sheet from a large pile.

'Explosion travelled that roadway, along which full tubs of coal were pulled by the endless wire-rope haulage system,' Jack positioned an oil lamp so they could see. Then fingers traced the route as he continued, 'That rope haulage is driven by the steam engine at the shaft bottom.' His fingers tapped on the drawing of the shaft bottom. Mr Garforth was looking and listening. 'It also travelled into the stables, but it died out on 't roadway along which water was carried to t' horses.' Jack continued. Still no comment from Mr Garforth. 'In t' incoming part of the system, where empty tubs travelled from the shaft bottom to near the coalface, the explosion died out quickly.' Mr Garforth continued to watch Jack's finger as it followed the explosions path. Jack stated, 'Explosion did not travel onto any o' working faces.' Jack marshalled his thoughts, then as he pointed to various spots on the plans, said, 'In each case it died out a few yards from t' coalface, here and there.'

'What was different about those areas, Jack?' Mr Garforth broke his silence.

'Roadways were thick with stone chippings, because they had been enlarged to allow ponies access to collect full tubs,' he answered

'Stone chippings. Carry on Jack.'

'Then it travelled along this area,' Jack continued as he picked up the last sheet of paper, 'That's where tubs are disconnected from ponies and lashed by means of them short, hooked chains onto endless haulage rope.' Mr Garforth's finger once again traced the route then he turned to Jack,

'What's your conclusions, is it gas?'

Jack stood up scratched his chin for a while, then said, 'Yes, we have gas, but the system you had installed o' drilling into likely areas of rock, then installing pipes and a tap to control the gas leakage, has always worked well. So possibilities of it being a gas explosion are very small.' Mr Garforth turned and sat down on the wooden bench as Jack paused. 'On the haulage road to the pit bottom, there was a thick layer of coal-dust. This had seeped out o' tubs as coal lumps rubbed against each other.' Again he paused while he pictured the scene. 'When we walked into the pit ont' haulage road, I can remember the sound we made as our clogs stirred up t' dust. After the explosion when I walked that road there was no dust.'

'What are you saying?' Mr Garforth asked.

'In the roadways where there had been no dust, where it was wet or stone chippings covered the dust, the explosion petered out.' He looked straight at Mr Garforth and added, 'It was coal-dust that caused explosion.' Mr Garforth looked at the plans then said,

'So you think it's coal-dust that caused the explosion, do you? You realise its never been given as a cause before?'

He was not certain but he could see no other reason,

'Yes,' he replied. With a smiling face, Mr Garforth said, 'Good man. You have confirmed what I suspected. Now we will face Mr Wardell together and convince him.' Mr Garforth walked over and shook his hand, then added as he made his way out, 'Make sure your notes are available for him.' Jack wrote down in his large handwriting his statement and the reasons, and then presented it to Harold for Mr Garforth the following morning.

Mr Wardell, the Mines Inspector, arrived later in the morning. He had read Jack's report when he was summoned.

'There must have been other places that were covered in stone chippings?' Mr Wardell asked.

'Yes sir,' answered Jack, 'but they were covered by coal-dust.'

'The explosion travelled over them, is that what you're saying?'

'Yes sir. The explosion only stopped at the stone chippings when there was no layer o' dust covering them.' Mr Wardell then turned to Mr Garforth and said, 'I am convinced.'

At the inquest Mr Wardell reported his verdict.

'The whole of the workmen killed, except the two deputies, met their deaths from an explosion of coal-dust. The two deputies, Deakin and Lomax, were suffocated by the stoppage of ventilation, a consequence of the explosion.' He continued after a short pause, 'I cannot speak too highly of the way in which one and all connected with the slow, tedious operation, which ultimately resulted in the fires being extinguished, the bodies recovered and the colliery restored to its working condition, acted.' He allowed time for the reporters, then continued, 'The official staff at the colliery and the bands of workmen all worked like true heroes.'

The newspaper headlines read:

COAL-DUST GIVEN BY THE MINES' INSPECTOR MR WARDELL AS THE CAUSE OF THE EXPLOSION AT THE WEST RIDING COLLIERY AT ALTOFTS IN YORKSHIRE

They all gave a long account of why that verdict had been reached. The mining world ridiculed the decision, and when Mr Garforth tried to explain why that decision had been reached, they also ridiculed him

In June 1887 the headline in the paper read:

COAL-DUST EXPLOSION AT GLAMORGAN'S NATIONAL YNYSHIE PIT

The mines' inspectors believed him, stating that the incident on 18 February 1887 at the colliery was due to an explosion of gas caused by shot-firing, and then accelerated by coal-dust.'

To cut down the risk of another coal-dust explosion at the mine, he insisted his men use gelatine instead of gunpowder. However they had to pay for the gelatine; it was part of their agreement and it was more expensive. With the coal trade once again in a depression, Mr Garforth could not afford to pay them more. They soon came out on strike.

William noted the colliery school had closed for the summer holidays as he drove his trap through a barrage of abuse at the colliery entrance. Again Mr Pickard, now Normanton's MP, visited Mr Garforth. He agreed with the introduction of gelatine instead of gunpowder, but argued it would cost the men more money. Once again Mr Garforth said he could not afford to pay them more.

Meanwhile he had been asked to give evidence at a Royal Commission on Mine Safety, and as part of his preparation, Mr Garforth hired an artist to paint what a speck of coal dust looked like under a microscope. Then he took photographs of the drawings, showing the glass negatives to the Royal Commission using a magic lantern projector powered by a candle. In the workshops almost every month, men seemed to be experimenting with each new discovery Mr Garforth read about. The use of the newly discovered electric power was always being tested, particularly the portable type that could replace oil lamps.

When the need for coal increased, he agreed to pay the men to use gelatine in places where there was most risk and they returned to work.

The cold morning rain swept into Jack's office as Mr Garforth entered. He sat on the wooden bench looking at the table for a while, then looked up.

'Right Jack, we both believe it was coal-dust, so even if no-one else does, we can make sure it never happens again at this colliery.' Jack waited. 'We will eliminate the cause of the explosion. You start by arranging to change all the tubs, and ensure the new ones don't leak dust. I'll look at the problem when the dust comes to the surface.' Almost as an afterthought he added, 'Also, make me a list of the equipment and other requirements used during our recovery of the colliery after the explosion.'

Later, after a quick look through the list Jack had given him, he said, 'Good. Now, make your recommendations for the minimum requirements.'

Two days later Jack presented his report. Mr Garforth glanced through it, then said, 'While I am studying this, you go out and ensure we have everything you have listed. If not, tell Mr Bramley he has my permission to purchase it.'

For the following months Jack was busy carrying out his requests. Meanwhile, Mr Garforth consulted with mining engineers in Britain and Europe, paying visits to see the latest technological innovations. Eventually Mr Garforth published a book entitled *The Recovery of a Coal Mine after an Explosion*. Explosions were happening almost every week, and any colliery that showed an interest was sent a free copy of the book. The Mining Society itself politely listened to him as he read them the book, but to prepare for an explosion which might not occur would cost money; lives were cheaper.

Meanwhile, William and Mary's family grew by another daughter, Christina.

One day, Jack and Mr Garforth were walking underground, beside a moving wire haulage rope. Suddenly a strand of metal wire that had come loose from the rope hit both of them across the back of their legs. Mr Garforth believed that once a safety hazard had occurred, it should not just be rectified, but its cause should also be eliminated.

'That's dangerous Jack,' Mr Garforth said, as they both looked at the red weals it had raised, 'I'm going back to the surface, to arrange for a piece of that haulage rope and every other wire rope, plus samples of the small lashing chains we use, to be sent to the workshops.' As an afterthought, he added, 'Jack, can you give me a report on any other damaged ones.' Jack waited until Mr Garforth's light had disappeared into the blackness, then muttered to himself, 'Thanks. We have over twenty miles of haulage-rope roadways with lashing chains scattered beside them, and throughout the rest of the pit, that means I've a lot of walking to do.'

Mr Garforth looked at different types of old wire rope with his microscope, and set up a breaking-strain testing procedure for wire ropes and chains, with Bramley his engineer. They got in touch with the makers, Messrs E. Baylie & Company, Chain-Makers, based in Stowbridge. As information was gathered, William collated it, once again publishing his findings in the *Institute of Mining Engineers'* journal.

He walked into Jack's office one day with a pile of books, 'Jack, I've another project for you. How can we increase men's safety when involved in a rescue?' He put the books on Jack's desk. 'I've read all these, now you read them and get all the useful information. I will write up my notes, and then we'll discuss it.'

They spent many hours at work and at Halesfield House, discussing the problems. 'We have to introduce training for rescue men, and also some kind of breathing apparatus, but which?' Mr Garforth said one day, 'English, German, French or American?' Mining journals were thrown on the table in front of Jack, and they discussed the merits of the breathing apparatus mentioned in them.

'Get me a ticket, Jack, I'll go home and have a word with Mary,' Mr Garforth would say if he considered it worth further inspection. A first class train and maybe a boat ticket were purchased, and off he would go. He became well known and respected in the German and French mining industry.

He travelled to local collieries that were trying out new types of mining equipment and made detailed notes about everything. If he liked what he saw, he would arrange for someone to stay and work with the equipment, and afterwards would study the report. On one visit he saw an electrically-driven coal cutting machine using a large revolving arm to undercut the coal; another had a large wheel fitted sideways. Both had small, hardened, steel pick-heads attached. He believed that by getting the rock above the removed coal, i.e. using a coal-cutting machine instead of men to drop in a straight line, they could control the fall and the men could work in safer conditions.

The men did not like it, believing he was trying to do them out of a job. They had not liked it when he had introduced the safety lamp instead of candles, and had gone on another strike because the lamp did not give out as much light as a candle. He also successfully introduced an electric fan, as this increased the ventilation and did not effect a man's earning capacity.

Like everything Mr Garforth tried out, notes were made of their strengths and weaknesses. Even when the bearings wore out, he analysed them in the workshop, and once again all his research was brought to the mining engineers' attention. By March 1889, even though he still had not convinced them about the cause of the explosion and the need to be prepared, they respected him. He had been elected as Vice-President of the Midland Institute of Mining Engineers, and was one of their representatives at the National Institute. At that time, almost one million men were employed in the mining industry. Over 100,000 were on strike, to try and force the colliery owners to recognise their union; it was June before the men of the West Riding Colliery joined the strike.

On 10 August 1889, a compromise was reached and the strike was over.

The population in the village over the last fifty years had grown from 400 to 4,000, and in the last year, eighty-six children under the age of one year had died. However, a supply of clean pipe water had now reached the village, and the first of the gas lamps at the side of the roads had been installed. Travelling theatre groups visited regularly; classes had started to teach the adults to read and write. Something new seemed to be happening almost every week in the village, and at the colliery. What would happen next?

eight

1890-1895
THE FIGHT TO GAIN RESPECT

1890. Llanerch Colliery, Monmouthshire, 176 killed. Morfa Colliery, Glamorganshire, eighty-seven killed.

The mining explosions continued, and William Garforth was invited to give evidence to the Royal Commission appointed to study the question of coal-dust explosions. After giving his evidence, he suggested that experiments should be made on different types of coal and dirt dust, as he believed mixing dirt-dust with coal-dust could be the means of preventing explosions.

On the long, tiring journey home, William reviewed, over and over, the next step in his long fight to improve the safety and profit of coal mines. Their growing family of daughters, Margaret, Helen, Frances, Christina and Elizabeth, had not stopped his wife Mary in her crusade. She chaired many committees at their home, organising, amongst other things, outings to the coast. She also raised funds to build an institute where the young men could go, rather than to the local public houses. At very young ages, she also involved her daughters.

The local Normanton Board laughed when they received a letter from Nicholson and Jenny's of Newcastle, asking the board if they were interested in lighting the streets with electric lights. Some of the councillors did not even know what an electric light was!

Over 2000 trains a day were now using Normanton station, and a constant stream of horses and traps, wagonettes and railway-drays coming from the station in deep mud were making it dangerous, as people could not get out of the way.

On the 13 June 1891, Mr Garforth answered questions put to him by Normanton Council on the need for educational facilities. At that time, besides his mining interests, he was Chairman of the Altofts Council, Chairman of the Wakefield Technical Education Board, and had been invited to join the Leeds and Barnsley Education Boards. His knowledge of the educational system, both in Britain and Europe, was unmatched, and he always volunteered his services to local towns

1892. Great Western Colliery, Glamorganshire, sixty-three killed. Park Slip Colliery, 112 killed. Mr Garforth was elected President of the Midland Mining Institute, and gave the following presidential address to over 500 mining engineers,

'The increasing depth of our mines is causing many problems. We need improved ventilation, and as better seams are worked out, washing and screening coal have to be considered.' He gave detailed explanations and answers. He suggested prizes should be offered to all learned engineers who were prepared to contribute papers on practical ways to make mines safer and more profitable.

They were already installing an automated washing plant at the colliery, the first in Britain.

In the workshops, they were experimenting with new coal cutters.

More stone-flagged pavements were being laid in the village. However, the roadways in winter were still almost impassable for the horses and carts, due to the deep mud. Almost 4,000 people now lived in the colliery housing. It was still separate from the rest of Altofts, but other speculative housing was being built on any available site. At the other end of the village, nearest to Normanton, better houses were being built for the management personnel of the various companies that operated from Normanton station.

'How's the first aid class coming along?' Mr Garforth asked as he stepped down from the carriage. Jack had now joined hundreds of men at the colliery, giving up their time to learn first aid.

'Fine,' he answered.

'Good, it's about time we tried to improve standards. Jack, you're going to help me organise a competition,' he said as he followed Jack into his office. 'This is a rough draft of an invitation. Send it to all the organisations we know who are training men in first aid. I will get in touch with the press and arrange a venue,' said Mr Garforth as he gave Jack a pile of notes, 'I'll award a silver rose bowl to the best team,' he added, as he donned his top hat and left the office.

A few months later, what had been a river of people in the early morning, from the nearby Wakefield, Normanton, Altofts, Featherstone and Castleford, had dried to a trickle by the time the warm sun had chased away the early morning dew. Over 10,000 now filled the higher slopes of the common land called Heath Common, between Normanton and Wakefield.

Below them as it sloped towards Wakefield, in a roped off area, the flames slowly died from a purpose built and fired wooden building. While they watched, adults and even children, all with blackened faces, moved into the building. Some lay on the ground as if dead, others cried out as if in agony. Meanwhile, the first of the first aid teams was moving into the ruins. Two doctors watched their progress. Shortly after, the first of the traders arrived, pushing their cycles or hand carts, with baskets or trays on the front, shouting, 'Boiled sweets, ice cream, pies, pikelets, etc', all freshly made in their kitchens the night before. Other traders sold pop, beer, etc., from their horse-drawn wagons. Children enjoyed a miniature round-a-bout, or listened to the barrel-organ man with his monkey. All announced their wares in different ways and it all added to the excited hum of the crowd. All day long the admiring adult spectators applauded the skills of the 700 first aid workers who took part in the exhibition, while the younger ones made new friends.

In the cool of the evening, Mr Garforth shook hands with the doctors as they climbed into their carriages after a long and tiring day, and then moved over to where Jack and his team were tidying up.

'If that saves one life, all our work will have been worthwhile,' he said, as they looked towards him. He looked around and said 'All your families have gone, good, I am escorting my family home, and then I will be back with a wagonette.'

On his return, Jack was invited to sit on the front seat with Mr Garforth and George Griffin, the horse driver. He sat in silence as they plodded up the hill that overlooks Normanton, and as they reached the top and looked down on the smoke haze that blanketed the town, Jack asked,

'Why do you help the Normanton board, when you don't agree with their politics?' He was aware that the Normanton Board members were practically all miners.

'The West Riding Colliery pays its rates to Altofts. Normanton have several large collieries just outside the town's boundaries, while most of the men live in the town.' Jack waited, as Mr Garforth continued, 'Without the income of rates from those collieries, the town council have struggled and got into debt and they, like Altofts, are striving to improve the living conditions for their inhabitants, and remember, many of our workmen live there. They've also asked me to be the chairman of the new Grammar School they are hoping to build, and you know of my interest in improving our workmen's education, so I've accepted.'

Mrs Garforth was also very busy. She encouraged the village to raise money and purchase an old malt-kiln on the main road so that the people of the village could meet and organise events.

In 1893 at Thornhill Colliery, Yorkshire, 139 men were killed. At Great Western Colliery, Glamorganshire, another sixty-three were killed. 139 men and boys were also killed in an underground explosion in nearby Dewsbury.

Once again the summer brought a depression in the coal trade. Throughout Great Britain, 1 million miners were on short time; many who were laid off and without work were soon reduced to starvation levels. The collieries around Altofts and Normanton were amongst the most efficient in the world, yet even here, they were only working one or two days per week.

Mr Garforth also looked after his family's interests with a colliery he had helped to finance in Ashton-Under-Lyne. This was the Ashton Moss Colliery, sunk to a depth of 932 yards, which was the deepest colliery in Great Britain. As its Managing Director, he regularly travelled over to consult with its manager, and was re-elected President of the Midlands Mining Engineers. For his presidential speech he updated his notes on recovering and rendering assistance in collieries after an explosion. He was also elected as Vice Chairman of the National Federation of Mining Engineers.

In early September 1893, with feelings running high throughout the country, the mine owners demanded a 25 per cent reduction in the miners' wages. The few men who worked faced a barrage of abuse wherever they went. Mine owners petitioned the government for protection, and sixty men of the 17th Lancers and their officers were stationed in nearby Normanton as tension grew.

Over 250,000 miners came out on strike; 6000 gathered in the fields between Altofts and Normanton to hear one of the miners' leaders, Mr Cowey.

Mr Garforth wrote to Mr Pickard MP, now President of the National Federation of Great Britain's Miners, saying that he was anxious that no unnecessary friction should be caused in the village, as they had worked amicably together for many years.

Mary, meanwhile, had presented William with his sixth daughter, Stephanie, and she was also paying for more food to be delivered to the colliery school for the miners' children's meals.

The families, rather than cause trouble in the village, walked to a mass meeting at the nearby Ackton Hall Colliery at Featherstone on 7 September, 1893. They joined thousands of men, women and children, many in ragged clothes, all with worn, gaunt faces trudging the many tracks. Like the spokes of a wheel, all led to Ackton Hall Colliery. The message had gone out by word of mouth that the owner was going to move coal that had been stocked on the colliery surface.

By early evening, as the tension mounted, the now several thousand-large crowd spilled into the colliery yard, and with grim determination showing on their faces, shoulder to shoulder the men edged forward as a barricade.

Wakefield, 4th Sept 1893.
9 p m.

Sir,
I have to inform you that as this City is surrounded at very short distances by Collieries and having regard to the disturbances and riots which are daily taking place in districts not far removed from this City and the grave apprehension that the Collieries adjacent to the City will be visited by the Rioters at any hour it was thought desirable that a meeting of the Justices acting for the City should be held to consider the desirability of steps being taken to preserve order within the City and that a meeting of the City Justices was so held to night at which they were unanimously of opinion that it was desirable that a request should be made to you to order a Troop of Cavalry to be held in readiness so that upon a telegram being sent to you to morrow, after a consultation between the Justices of this City and Justices of the West Riding, that the Troop should be ordered to Wakefield, you will order their immediate transport to this City to assist the Police, if required, in the protection of the persons and property of the Inhabitants.
On behalf therefore of the City Justices I beg to make such request accordingly.

I am, Sir,
Your obedient Servant,
John S Booth "Mayor".

The Right Honble
The Secretary of State.

The local Mayor was expecting trouble.

'Fix bayonets!' The order stopped the forward movement, and all was silent as twenty-eight South Staffordshire Regiment soldiers' bayonets flashed in the sun as they were fixed to rifles. They had been sent from nearby Bradford to help deal with the crowds.

'Please disperse,' the colliery owner shouted. An angry growl went through the crowd and slowly built to a crescendo. Seven weeks on strike and prior to that only one or two days work a week; they had no money and their families were starving. The horses of the colliery owner and local magistrate, sensing the tension, became nervous; the soldiers waited. A gap appeared in the bayonets as Mr Hartley, the magistrate, advanced with Captain Barker towards the crowd. The captain held up an oil lamp in the now quiet, and Mr Hartley used its light to read the Riot Act.

One man's cry broke the silence which followed this reading. 'Our families need food!' The cry was taken up by the crowd, who surged forward, closing the space between them and the soldiers.

'Shoulder arms!' The surge checked as, with precision, the soldiers moved their rifles. 'First section, prepare to fire!' Once again the crowd stilled, the bayonets flashing as the rifles moved through ninety degrees and pointed towards them. In almost total silence, the click was heard as rounds were fed into rifle barrels. Under pressure from the large numbers who could not see the soldiers, the crowd pushed forward.

'Fire!' The bullets blasted into the tightly packed crowd. 'Reload!' The clicking sound could be heard through the gun smoke. Momentarily as the smoke drifted over the front of the crowd, time stood still. They moved forward again. 'Second section, prepare to fire.'

'Fire!' The shots rang out. 'Reload!' Empty rounds hit the floor. There was a click as new rounds entered the rifle barrels.

Women and children began crying; men and women dropped beside the bodies that had slowly crumbled to the ground. A slight movement backwards became a race, and more were injured by the crowd who were now desperate to get away. Small groups, braver than the rest, gathered around the two dead men, James Gibbs and James Duggan, and the many more wounded. Miners used their newly acquired first aid skills to staunch the bleeding. The soldiers, still prepared to fire, watched as they were carried away. James Gibbs was taken home to Loscoe, near Normanton. James Duggan was taken to his home in Featherstone. Five of the wounded also returned to Normanton.

When the mild winds of September turned cold in November, the starving miners accepted the owners' terms and returned to work.

In 1894 at Albion Colliery, Glamorganshire, 278 workers were killed.

Mr Garforth now had a very high reputation in the coal fields of Europe, even though in Great Britain they still did not totally believe his theories. He was invited to the opening of an experimental mining gallery set up in Germany. In several five-metre long iron tunnels, they were carrying out investigations to discover the causes of underground explosions.

Meanwhile, in Germany, France, America and Great Britain, more major explosions rocked the industry, and thousands were killed. In Great Britain, the mines inspectors stated that two were the result of explosions of gas and coal dust. The miners, most unable to read and write and with very little contact with the outside world, remained unaware of the dangers.

nine

1895-1900
A Safer Working
Environment

A noisy crowd, shuffling together in the biting wind, suddenly hushed. All eyes stared in wonder. A train squealed to a halt in a cloud of steam. After a moment, the passengers started to alighted. Shining out through the smoke from the West Riding Colliery chimneys was a giant crown, outlined in electric lights on the cage headgear.

'How have they managed to do that?' Harriet Burr whispered. Bill, with their daughter Lucy on his shoulders, did not hear her the first time. However, when she pulled his arm and asked again, without a movement of his eyes, answered,

'It's all done by pulling a lever down.'

'What do you mean Bill?' Harriet asked.

'I'll explain later,' Bill replied. He had seen an electric light in the colliery workshop, but it was unbelievable to him that a spark caused by pulling down a lever could travel along a piece of wire and shine in a glass bowl. Somehow, those same sparks had been multiplied to fill the hundreds of glass bulbs that formed the giant crown. That announced to the world that Altofts knew how to celebrate Queen Victoria's Diamond Jubilee year.

Below the crown, Normanton's fire brigade stood, their new steel helmets reflecting the lights as two stepped forward and demonstrated their new wonder, a telescopic ladder. Another passenger train issuing clouds of steam came to a halt, breaking the silence. Bill looked round, the steam distorting the light, making the passengers alighting from it look like aliens from another world. He squeezed Harriet's hand as he turned back. For a long time, hardly a voice was heard in the now massive crowd. Then a train's whistle recalling its passengers stirred everyone into a frenzy of sound and activity.

Many of the crowd had never seen electric lights. Bill and Harriet stayed for another thirty minutes, then slowly made their way home, looking back every few yards, still not able to believe their eyes. It had been a marvellous day; tea, cakes and a presentation mug for the kids, a football match and then this spectacular evening.

The next day, the colliers voted to go on strike rather than allow the introduction of coal-cutting machines.

Over the hum of people working in the parish rooms was a buzz of excitement. Men were making furniture and toys, and chopping up the left over wood for the fire.

'Any news Bill?' a voice called through the blue haze of pipe smoke.

'Union are in with Mr Garforth now. I hope they settle this strike,' he answered, as all the noise ceased.

An early coal-cutter being checked in the colliery workshop prior to going underground.

'Aye, and get rid of those coal-cutting machines. They do away with men's jobs. We'll not go back until Garforth backs down!' someone shouted across the room, to a mutter of approval.

'So what happens if they start bringing in men from other pits?' a woman's voice cut through the subdued muttering from another room.

'Blacklegs, we'll sort them out,' she was answered.

'Aye, but we'll all be evicted,' Bill said, 'and I know what it's like looking for somewhere to live.' He remembered going to a two-bedroom house, which had eleven children sleeping in one bedroom, and they'd still offered him and Harriet a room. Many others they had called at were almost as crowded. 'I hope they settle it soon,' Bill thought of the three-bedroom colliery-owned house they now had. 'My Harriet's not happy with me being on strike,' Ribald's comments followed.

Strikes were a regular feature of the village as the miners fought the introduction of any new machinery that they thought would result in men losing their jobs.

Sixty-three men and boys were killed in a coal-dust explosion at the nearby Micklefield colliery.

In nearby Normanton the council were sending out letters to the local, major property owners, stating many of their houses had to be declared unfit for human habitation for the following reasons:

1. There is no sewerage system to allow the slops from the houses to drain away.
2. No running water is available, either in the properties or within 300 yards.
3. There are not sufficient toilets.

They were available to discuss these and other problems, such as provision of a causeway and some maintenance of the roads to keep the mud down.

After nine months the strike ended. With little food, children were dying of scarlet fever. Mr Garforth took out the coal cutters, saying they were not reliable.

Against this background of unrest, Mr Garforth and his engineering staff were constantly evaluating new mining apparatus as it came on the market. Some they tried to improve. The latest being tested was a Sutcliffe coal-cutting machine. All of the other coal-cutters they tried had constantly broken down. Once again, the cutting wheel was enlarged and this time they strengthened all the parts which they had found from experience had broken down previously.

'Jack, now they've finished the improvements to the Sutcliffe coal-cutter, can you take it down and try it out?' Mr Garforth asked.

'What about the coal hewers?' Jack replied. 'They won't like it.'

'Tell them I'll work out a bonus with them so that everyone gets more money if it works okay,' Mr Garforth replied.

'Can we stop the haulage rope system to lash on the coal-cutter parts?' Jack asked. Lashing was a system of using small chains with hooks on to attach small wagons onto a moving wire rope haulage.

'Why?' asked Mr Garforth. 'We don't when lashing on coal tubs'

'The parts of the coal-cutter on the flat wagons make it awkward to lash them on while they're moving.' Jack held up his left hand.

'Alright,' Mr Garforth replied as he noticed the missing tip of Jack's finger, 'but do it on the afternoon shift when we're not producing any coal, and also run in the pipes we'll need to convey the compressed air from the steam boilers to provide power for the coal-cutter.'

Jack arranged for the coal cutter to be dismantled, each part slung under the cage and taken down the shaft. Then using tension chains, they were clamped on top of a flat wagon. A hook on a small chain was attached to the wagon and the other end lashed around the endless rope haulage, ready for the afternoon shift to transport to the face.

'How is it going?' Jack called the next morning as he peered into the flickering light supplied by an oil lamp. His hand removed one very sharp piece of rock from the bed of coal and small rocks he lay on.

'Should be finished in about an hour,' the reply echoed back from between the short stubby wooden props that supported the coal roof.

'Good. When you've finished, help anyone who is behind. We must ensure that the coal face is straight for at least 100 yards so the coal-cutter can cut. I'll make sure you get paid for the extra work.' The click of an iron pick on rock answered him, and in a crouching walk, he shuffled on to the next team to check its progress.

'Give us a hand, Jack?' Jack pulled on the rope that was attached to the stubby props which supported the coal and with several sharp cracks, the coal dropped. Even before the sound had died away, the team and Jack moved forward into the dust to start the process of loading the lumps of coal into a waiting tub.

'Jack, come and fire us some shots,' a voice echoed along the dust filled passage. Again in a crouching walk, he moved down the face line to where the removal of the props had not caused the coal to drop and crack. Small, deep holes had been drilled into the coal and packed with explosives. Jack pushed detonators into the explosive, then connected them up to the waiting cable. With his oil lamp held close to it, he ran the cable over his hand as he moved back, checking it. Crouching behind a coal tub, he connected the cable end to his firing battery.

'Fire!' he shouted to men, hidden behind any safe refuge. The blast burst on their ears.

For a short while, choking dust was swirling all around, along with the smell and taste of the explosion, and then slowly it drifted along the face-line, moved by the incoming ventilation.

Jack moved on. Fortunately, he only had to concentrate on about 100 yards, not the whole face line of over a mile. Soon black, sweating bodies gathered in the dim light around the last tub, full of coal lumps, ready to push it off the face and harness it to the waiting pit pony.

'We've just time to drag it onto the face,' Jack said, indicating the parts of the coal cutter that were gathering dust in the roadside, 'We'll leave it for the mechanics to reassemble on afternoon shift.' A mouse scurried away, leaving its tracks in the mixture of coal and stone dust as they laid small wooden props down. Muscles straining, the rope they had placed around a pulley tightened and the wood support it was fastened to groaned, but the rocks laid down over millions of years held it securely in place. Those same rocks moaned as the cutter, scraping the roof, was hauled slowly inch by inch over the small wooden props onto the face. 'Why are we installing another cutter?' someone asked as they walked the dark miles to the shaft bottom, occasionally moving into the side of the roadway when the rumble of iron wheels announced the approach of tubs, lashed to the endless rope haulage.

'Mr Garforth has promised that the men who run the coal cutters will be paid more, and it will eliminate the dangerous job of working under the coal,' Jack replied.

'We've heard those reasons before,' a voice came through the dust and the sound of clogs, 'and it's doing away with men's work.'

'Look, Mr Garforth says no-one will lose their job. What should happen is that all the hewers can concentrate on getting the coal out once the machine has undercut. More coal per team means more money.' Jack argued with them.

'You're a boss's man.' one said. They no longer trusted him.

Only Mr Garforth's promise of money to test the machine had settled the initial problem. The next morning, with the cutter connected up to the compressed air, then with its large cutting wheel cutting under the coal, it slowly moved along the face.

'We cut 50 yards with no serious problems.' Jack was able to report to Mr Garforth at the end of the shift.

'Stay with them for a couple of weeks, Jack, and keep me informed,' Mr Garforth replied.

The coal-cutter had many modifications, and the distance cut per day improved. This time they had a good machine. Mr Garforth eventually arranged to purchase Mr Sutcliffe's patent. He renamed the machine, in honour of the Queen, the Diamond Coal Cutter, then along with the changes, patented it.

'What's it like, this coal-cutter?' young Mary Prendergast, knowing of the tension in the village, asked her father James one day.

'It's like an iron man.' Mr Garforth, standing on the other side of a wall, heard James reply.

When Mr Garforth referred to the Diamond Coal Cutter after that, he called it the Iron Man and the name soon caught on. Now he had two problems: he required more machines to be built and he needed to persuade more hewers to use them.

At the colliery they had also discovered the Upper Beeston seam 80 yards below the Middleton Seam. The coal was workable, and they decided to call it the Diamond Seam in honour of the Queen's Jubilee.

The Diamond company coal-cutter, 'the Iron Man'. The Diamond Company, founded by W.E Garforth in 1897, continued to manufacture mining equipment for over 100 years. (By undercutting the coal, men no longer had to carry out this dangerous operation.)

In 1896, William made a recommendation to Barnsley Council that they should establish a Mining and Engineering College, which was heatedly debated. The local landowners and mine management argued that there was little point in technical training when a lot of workers have not got even an elementary education. The British government stated that it attached no great value to scientific training, arguing that young men would prefer to play sport. Mr Garforth was frustrated that they couldn't see that better educated workmen would mean fewer accidents, more output and more profit.

The February snow of 1897 was still on the ground as the coal hewers of the West Riding Colliery, specifically the Haigmore coal seam, gathered in the Miners Arms public house.

'Garforth's going to introduce coal-cutters into our seam,' the chairman said, 'what are we going to do about it?'

'Strike!' came the unanimous answer.

'Why?' Jack asked Bill after the meeting, as they sat in his hut at the bottom of his yard, 'it's safer, the men working the Iron Man are getting paid 7s 9d, that is good money.'

'Aye it's good for them working the Iron Man, Mr Garforth, but what about rest? Even them that are working with the machine say it's dangerous, they can't hear anything but the cutter.' Bill stopped and looked at Jack. 'You know as well as I do, if we can't hear the sound the roof makes as it starts to crack, we can't get out of the way when it drops.' Jack had no answer to that.

Mr Garforth constantly proposed to the many mining organisations the creation of mines' rescue stations. He believed that groups of collieries should combine to finance a central rescue station, which could train men and store equipment. It caused many heated debates, and one mining inspector commented that when he visited collieries after an explosion, which were working on the basis of the Garforth rules, they seemed organised, whilst at others there was lack of discipline, with the result that the officials made frantic efforts, only to be exhausted and unable to tell the strangers, who had volunteered to help from other collieries, what the situation was.

The president of that same organisation stated that in his view, he did not think it was good policy to anticipate an explosion, it was better to wait until the Mines Inspector arrived to take over.

William at times despaired, but with Mary's full encouragement, William carried on the fight to improve the safety of Great Britain's mines.

However, many were beginning to take notice. He was elected President of the Midland Mining Institute, and a Vice-President on the National Body, and they accepted his offer to test rescue apparatus. Another Royal Commission, this time on the use of electricity underground, asked him to give evidence. However, many owners still did not see the need to spend money on safety underground, refusing to spend money on preventing coal-dust explosions, particularly if they had never had one, or didn't believe it was a proven cause.

That same year, 1897, Mr Garforth addressed the International Life Saving Congress, at Frankfurt in Germany. After a long speech outlining many safety features he had introduced into his collieries, he finished by saying, 'For every 675 miners employed, one will lose his life.'

Afterwards, the President of the Congress said, 'This silver medal is our highest honour, and we award it to Mr Garforth in recognition of the work he has done towards the safety of the mining industry.'

Meanwhile at Altofts, the miners in one coal seam were once again on strike over the introduction of coal-cutters. Mr Garforth told them it made it safer to work the coal, and more profitable, that meaning he could pay better wages. They said it did away with men's jobs.

The Iron Man coal-cutters were working in other seams at the colliery and at other collieries in Britain, Germany and France.

Meanwhile throughout the country, Mines Inspectors and influential mining engineers were saying they should use the Garforth rules for preparing and recovering a colliery after an explosion.

William accepted the chairmanship of the governors of the new grammar school in Normanton.

The coal-cutters were doing their work well; potential buyers came to see them in action and it soon became obvious a separate workshop had to be built to cope with the demand. As a result, Mr Garforth formed the Diamond Coal-Cutter Company, and a small cottage on Altofts Road, Normanton was used as the offices. It was, after all, Queen Victoria's Diamond Jubilee when they started manufacturing. Orders for the coal-cutter arrived from all over the country and the continent of Europe, and were passed on to the new office. Soon, the four Diamond Coal-Cutter Company employees could not cope with the demand, and a new factory was built on Stennard Island, a small piece of land between the river and canal at nearby Wakefield. Here, Mr Garforth replaced the compressed air used in the cutters by electricity. These coal cutters were also successful.

Another new device, hydraulic props, were introduced at the colliery. 'Do we have to recover these hydraulic iron props?' Jack was asked one morning, as he paid one of his periodic visits to a face. The voice came from a coal-hewer who laid full length, trying to pull one from under a large rock.

'Yes, they are very expensive.' Jack replied. Only the previous day, he had been lectured by Mr Garforth on the cost and how many had been lost.

'Why do we have to use them then?' another voice asked in the semi-dark.

'You are safer with Hydraulic props,'

'Safer! With wooden props we can hear them creaking when ' weight o' rocks over us start to move, they warn us so we can set more support or get out o' way.'

'Do you call this safe?' The other hewer had just managed to remove the hydraulic prop back to safety when a sharp crack and whoosh of air announced the old face line had dropped, closing the space where he had just laid, 'We would have just left wooden props.'

'I'll see Mr Garforth when I get out,' Jack answered.

That year the Chinese visited the colliery. William Garforth also became a Fellow of the National Geological Society of London.

It was a long, harsh winter in 1898. The latest strike, over the introduction of coal-cutting machinery, lasted over six months. Miners, like Bill, were refusing to allow the coal-cutter, known as the Iron Man, into their coal seam. Children no longer played with smiling happy faces, they just stood around with pale faces and thin bodies. Each working miner in the colliery village helped his striking neighbours. The summer sun of 1899 brought little relief to the starving families. Police and magistrates were evicting tenants who had not paid their rent.

The magistrate waited until the quiet crowd stopped their forward movement. 'This eviction order is dated 6 May 1899, he read from his horse. 'It requires the men and their families, who have been in dispute with the West Riding Colliery Company for the past eighty-six weeks over the introduction of coal-cutters, should vacate their colliery houses within the next twenty-one days.' Harriet squeezed Bill's hand as he read out a list of six families, then relaxed. They were saved for the moment.

'We have lived here twenty years,' a mother cried out.

'We have lived here thirty years,' another commented, shoulders slumped.

'You have not paid any rent for eighteen months. Mr Garforth has been extremely generous in allowing you to stay so long in my estimation.' The magistrate paused and looked at the police constables, then at the now-silent crowd. 'You have twenty-one days. You must leave your furniture, as this will be sold to pay off your debts when the bailiffs move in.' Then he turned and rode away.

'What do we do?' said one of the evicted miners, turning to Mr Pickard who was standing nearby.

'You are still getting supplies and money from the Miners Association. Keep up your strength lads, we will win in the end,' he replied.

'I would be better off fighting the Boers, at least my wife would be able to feed the kids. And I'm not the only one who feels that way.' the man said as he turned to his sobbing wife.

Families were already occupying wooden huts they had built in the nearby Fox Holes Lane after they had been evicted. Army tents joined the huts in the adjoining fields, waiting for more families. William Garforth had offered to suspend eviction orders if only they would compromise, but their stubborn pride made it all the harder.

Mary Garforth visited those wooden huts with fresh provisions and, after much argument with their wives, the striking miners accepted that the wife of their enemy could provide them with food and clothes

Eventually the coal trade picked up again, and Mr. Garforth was able to offer a rise of *6d* which meant they earned 8s a day when they used the coal cutters.

ten

1900-1905
THE RESCUE APPARATUS

From the safety of a cold, damp brick wall on a February morning in 1900, a young boy looked around the dimly-lit colliery yard. He knew he should not be here, but his father was away fighting the Boers. Eyes twinkling, enjoying the challenge, he looked towards the Silkstone Shaft. Intermittent lights were coming under the door of the room beneath the headgear. He looked up as a boiler let off steam, then, mud splattering up his white legs, his boots clattering on the cobbles, he ran across towards the lights. Ears listening and chest heaving, he crouched near the door. No-one shouted as he peered under the door. Then quickly he jerked it back. Steeling himself, he looked again, and as his eyes adjusted to the light he noticed a strange creature. It had enormous protruding eyes and waving tentacles, with lights on the ends of them and strange humps on its back. Another creature lumbered out of the smoke carrying a body. Over its enormous nose, flickers of fire danced in fierce eyes that looked towards him. He jumped away from the door and had only run 10 yards when he ran into a man. His mind raced as he looked up; it was one of the creatures.

'What do we have here?' a strange sounding voice asked. In panic, he fought to get out of its grip and once free, fled.

Meanwhile, Jack was just locking his office door, but as he turned in the light of the moon, he froze. A strange humped-back being approached. It was hairless with large round eyes and a trunk. He hear, a harsh grating sound as its lungs expanded, then, a strange gravelly voice said.

'Did I frighten you Jack? I've just come across some youngster who was taking a short cut through the pit yard,' came the harsh grating once again, 'he'll be talking about monsters for weeks. I've never seen anyone run so fast.'

'Is that you Bill?' Jack swallowed then remembered the self-breathing apparatus, and that Bill Burr had been one of the first to volunteer.

'Aye, didn't you recognise me?' Bill laughed.

'Well, what do you think of it?' Jack asked, now his breathing had returned to normal.

'It's heavy, and at times the air doesn't come through so I wouldn't trust it in gas.' The harsh grating sounded again as Bill breathed in. 'It's easy walking round here, but what would it be like in conditions underground.' Bill replied and walked off into the dark.

The next day, Jack reported Bill's comments to Mr Garforth. 'Right, so we have some problems with the apparatus we got from Ackton Hall colliery do we? We will try and sort them out ourselves. Tell Bill I agree with his comments about creating a training area to simulate underground conditions.'

The miners testing the rescue equipment.

Many hundreds volunteered to carry out Mines Rescue training. The training was carried out in their own time.

After twenty years as Chairman of the Altofts Council, Mr Garforth asked if he was able to retire, but was begged to stay on.

Mr Bramley, the colliery engineer, was asked by Mr Garforth if he could improve the self-breathing suits. In the workshop, they spread the contents of cylinders, pieces of pipes, and various leather straps on a work bench.

'They'll have to be strong to carry that' Jack said just as Mr Garforth entered.

'Jack, do you remember those discussions we used to have?' Mr Garforth asked.

'About how we could organise a rescue team?' Jack suggested.

'Yes I've been thinking about the problems,' he turned towards Mr Bramley. 'Samuel, you know that area under the Silkstone shaft, where they have been testing the equipment? It's not realistic. We need it to be like an underground roadway after an explosion, and need to be able to feed smoke in, and maybe later gas.' He paused for a while. 'Many times I have explained the benefits of mines' rescue training school and station to the mining world.' Jack waited knowing there was more to come, 'They are very interested, but no one is prepared to cover the cost, so we are going to create a school.' Decision made, he turned to Jack, 'Can you get some more volunteers to wear the apparatus?' As an afterthought he added, 'make sure their comments are recorded after each trial.'

'Yes,' Jack said instantly. Even after fourteen years, he knew many still remembered that helpless feeling after the explosion.

'Good, let me have their comments as soon as possible. I will see Dr Mackenzie and get him to work out some exercises to improve the fitness of the volunteers.' Mr Garforth paused, 'Samuel, see what you can do to improve the apparatus, and remember they have to be able to work in gas.'

'Ask Charlie Illingworth to come and see me?' Mr Bramley asked Jack as Mr Garforth left. Jack crossed the colliery yard. All around was hissing steam and the hum of metal shafts and long leather belts that transferred their power to different colliery equipment. Steam from a leaking boiler was spurting all around Charlie when Jack found him.

'Mr Bramley wants to see you.' he called.

'Right I'll be there as soon as I've turned pressure down,' Charlie answered. Jack was wearing the apparatus when Charlie arrived.

'Charlie, I've got a job for you. They're still having problems with this self-breathing apparatus, can you see if you can help?' Mr Bramley asked. Charlie looked at the apparatus, now strapped to a leather apron over Jack's head and shoulders. On his back was an oxygen cylinder, from which flexible pipes ran over his shoulders to a unit that fitted over his face, with two large glass eye pieces.

'I can't breathe,' Jack shouted.

'Is that better Jack?' said Mr Bramley as he adjusted the valve on the oxygen cylinder.

'Aye, but now I can't see, the glass is misting up.'

'Jack, we've got to get the air and oxygen mixture right. I'm going to turn it down a little at a time. We'll have to devise a procedure for this, okay?'

Eventually Mr Bramley said 'I'm not happy with this control valve. We've got to be able to adjust it better. Have a look at that first Charlie.'

However, even the highly experienced Charlie could not solve the problems, so the apparatus was sent back as not suitable. As other firms announced they had new self-

This was the first Mines Rescue training gallery in the world.

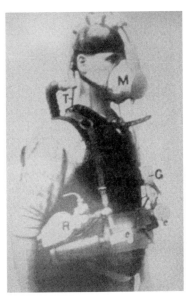

William Burr, dressed in the WEG Mines Rescue equipment.
(named after its inventor W.E. Garforth).

breathing apparatus, they were all tested. None proved totally reliable, even after Charlie
had once again tried his modifications.

Mr Garforth asked for unpaid volunteers to form a Mines Rescue team. Almost every
man at the pit volunteered.

All underwent medical, strength, and hearing checks, then tests to put them under
pressure. The ones who remained calm passed on to the next stage; each also had to qualify
in first aid. All through the early summer the tests continued.

One warm May night, the silence of the village was broken as a rocket with a long,
shiny tail screeched into the night sky. Six year old Nellie Welsby, excited and exhausted,
watched in wide-eyed wonder as many coloured stars dropped towards her. It was a new
century, and now Lord Roberts, with the help of local men, had achieved a victory over
the Boers at Pretoria. Her father John hugged his wife Harriet with one arm and held
Cyril, aged two, on his shoulders with the other. Born in Liverpool, he had lost both his
parents when he was three years old. Having no relatives, he was taken in by the Park
Road orphanage. He left aged 12 to work at the Stanley Colliery at Liversedge, where he
met and married Harriet Rothery. With a young baby, Nellie, they had moved to better-
paid work in Altofts five years previously. John was very proud that he had been selected
for the Mines Rescue team.

Meanwhile Mr Garforth and Mr Bramley had designed and built a new self-breathing
apparatus they called the WEG, after Mr Garforth's initials.

In the village hall, as Mrs Garforth rose from her seat on the stage to address the meeting,
the talking stopped.

'As you are aware we have now 255 men from the village fighting against the Boers. The
new volunteers need help, as the government only allows £4 16s for them to be kitted out.
We need more money.' Two volunteers, also on the stage, stood up at her signal. Their smart
bright red tunics, blue helmets and blue trousers with a bright red stripe contrasted sharply
with her dark brown dress. 'Mrs Craven,' she continued as they sat down, 'has kindly agreed

to read the latest letter she received from her son John.' As the applause died down, Mrs Craven stood up and, with tears in her eyes, she read,

'We spent thirty-two days on HMS *Pavonia* on dirty, dry, hard, stale biscuits and salted pork. The first day in South Africa, I was involved in a battle against 25,000 Boers. We lost 1,147 men, killed or wounded. We went at it for nine hours without food or drink. I was sunburnt something terrible.' She looked at the audience, 'The rest is personal.'

Instantly the audience stood up,

'God save our gracious Queen...,' they sang, as the box was passed down the rows, collecting bronze and occasional silver coins.

In the second week of June, the roads were bedecked with flags and the village rang with cheers as they celebrated. That night, as the dust in the colliery yard swirled around them, they watched the illumination of the Diamond Shaft headgear by hundreds of red and yellow lights, which spelt out 'V BOB', to celebrate Lord Roberts' victory and the relief of Mafeking.

Mr Garforth used the occasion to show a scale model of a bath house for the men, which was designed similarly to one he had seen in Germany.

On Sunday morning, the men from the colliery housing, in their best suits and boots, joined others from the area at the Miners Arms public house in Altofts.

'Well do we vote for this bath house or not?' the Chairman addressed the solid mass of miners.

'What's it going to cost us?' a different voice asked.

'¾ penny a week' said the Chairman. 'Any more questions? Okay, we will have that vote now. For?' Only the odd hand was raised. 'Against?' No count needed. 'Against has it. After all, everybody knows washing your back weakens it.'

In 1901 Queen Victoria died.

Altofts was described as a rich and prosperous village; water and sewage was connected throughout, and there were only seventy-nine back to back houses.

The Wagonette proprietors organised a time-table for a regular service into Normanton which would connect up with the electric tramway coming into Normanton from Castleford and Pontefract.

On 24 May 1901, as the Mine Rescue volunteers trained in the sun, eighty-one men and boys were killed in a coal dust explosion at the Universal Colliery, Senghenydd, Glamorganshire.

On a cool September afternoon, 140 members of the Midland Institute of Mining Engineers, in small groups, entered a sealed rubble-filled area under a colliery headgear, and having looked around, settled down on the seating provided outside.

'Gentlemen, let me welcome you to the West Riding Colliery School for training men in Mines Rescue, the first in the world.' William Garforth paused, 'After each training session, the men must fill in this book.' He picked up a brown leather-bound book, and placed it on a table. An engineer picked up the book; on the first page was written:

> Those who are our friends, tell us our faults. If you don't tell us of the faults of the apparatus, thus giving us a chance of improving it, then you are not our friends.

As a group of men moved into a roped-off area, William continued, 'We can feed smoke and gas into the simulated underground roadway.' He watched them prepare, then turned back to his audience.

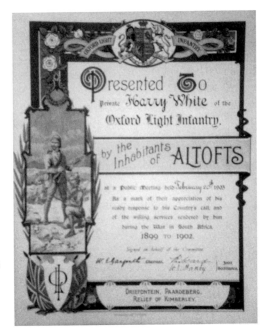

Very patriotic, Mr Garforth encouraged his men to support their country in time of trouble. This type of certificate was awarded to men who volunteered to fight in the Boer and the First World War. Note his signature.

'Gentlemen, this display is by men not wearing rescue apparatus, but it demonstrates what men in full rescue equipment will be undertaking in the smoke filled training centre you have just left.' Then he added, 'If we have any young men who would like to take part in the display, it can be arranged.' No-one replied so he carried on, 'The complete training takes two hours. However, time in the chamber has been limited to twenty minutes as there is no apparatus at the moment that can stand the conditions longer. I will give you a commentary.' Mr Garforth signalled to a waiting group of men who moved into the display area.

'First, to warm up the muscles, a period of lifting weights for example. They must be able to lift 56lb. a minimum of twenty-five times.' He paused while the men demonstrated. 'The rocks represent a fall of rock in the roadway and must be cleared.' Again he waited whilst men shovelled away many tons of rock. 'Speed is essential to save lives. However the rescue team's safety is paramount. The roof must be supported.' Men disappeared round the corner of a steam engine house, then reappeared carrying various wooden supports.

'Now, we need to seal off old workings from which gas may be leaking.' The men once again disappeared, only to reappear and move quickly along the length of the demonstration area carrying bricks mortar and water. Then they proceeded to build a small wall. 'Now they may have to crawl through a very restricted area; that is why we have used the pipe.' They watched as men crawled through a large pipe dragging wood and bricks.

'A body has been found. The dummy you see weighs 300lb. It could be a fellow rescuer wearing his apparatus.' The audience watched with fascination as the body was loaded onto a board, then man-handled through the pipes and over the rocks, back to the start. By this time, all the men carrying out the demonstration were breathing heavily. 'Thank you,' Mr Garforth turned back to his audience as the team walked away.

'Today, gentlemen, we are demonstrating a new apparatus which we have designed and is looking very promising. The chamber is now being filled with smoke.' The audience stood up and stretched. 'For thirty-five minutes, men wearing the new WEG rescue apparatus are to carry out some of the procedures you have just seen, in that smoke and rubble-filled area you previously left.' All available window space looking into the gallery was soon filled. William allowed the audience to look, then as some moved away to allow others to look, added, 'Only recently we spent seven weeks fighting an underground fire at the New Moss colliery, of which I am the Managing Director. The WEG was used successfully.' He paused and picked up a telephone and a reel of cable.

'The manager carrying out the first exploration used a telephone and was able to describe to me the conditions, so I was able to arrange for all material required to be sent down without him coming out of the mine.' he paused, 'All parties underground maintain communications by carrying forward telephones and cables. This saved lives when another party became unconscious from the effects of carbon monoxide given off by the fire; one managed to use the phone before he was also overcome, and a nearby party who had taken a WEG were able, by using it, to get them out.'

'How do they communicate when wearing the apparatus?' one asked.

'Each person carries a horn. The signals are, one is safe, two for danger,' William replied. As the spectators took turns looking through the smoke filled windows to watch what was happening inside, one asked,

'How many men have you in training?'

'We have whittled down the many hundreds of men who volunteered to thirty-five and remember, they undertake this training in their spare time.' William said, then glanced at his pocket watch. 'Gentlemen, the team will shortly be coming out. I have asked Dr Mackenzie,' a tall gentleman stood up and William continued, 'to explain what checks he carries out on the men when they have finished their training.'

Through the smoke, as it drifted out of the chamber, five men walked in full rescue apparatus. As they removed the apparatus from their sweating faces and shoulders, the doctor and his assistant moved in.

'Each one has his pulse and blood pressure checked, and they are reminded that they must breathe deeply to start the process of removing the carbon dioxide that has accumulated in their body.' said Dr Mackenzie.

'Please note gentlemen, how each apparatus is checked, and the amount of oxygen used is recorded.' added William. A couple of mining engineers moved over and lifted the rescue apparatus, then looked back into the chamber, 'They volunteer to do this training?' one stated, shaking his head. Another turned to his friend,

'Who is financing all this? It must be costing thousands of pounds?'

'Garforth,' was the reply.

These records still exist.

Local youngsters crowded the streets to watch the procession of carriages take them back to Normanton station. This was one of many similar visits and slowly as the apparatus was perfected, the length of time it could be worn was extended. For a short while, the proud village of Altofts became the centre of the mining world

In June 1902, as once again the shaft was illuminated to celebrate peace in South Africa, testing was being carried out on Meyer & Giersberg rescue apparatus, both built in Europe.

Eventually several collieries North of Sheffield did combine their efforts and built a Mines Rescue Station, based on Mr Garforth's design, at Tankersly. Mr Garforth was the guest of honour at the opening ceremonies. He gave this offer to the audience,

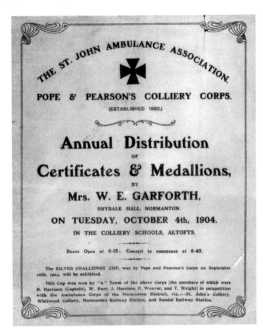

THE ST. JOHN AMBULANCE ASSOCIATION.

POPE & PEARSON'S COLLIERY CORPS.

(ESTABLISHED 1883.)

Annual Distribution

OF

Certificates & Medallions,

BY

Mrs. W. E. GARFORTH,

SNYDALE HALL, NORMANTON.

ON TUESDAY, OCTOBER 4th, 1904.

IN THE COLLIERY SCHOOLS, ALTOFTS.

Doors Open at 6-15; Concert to commence at 6-45.

The SILVER CHALLENGE CUP, won by Pope and Pearson's Corps on September 10th, 1904, will be exhibited.

This Cup was won by "A" Team of the above Corps (the members of which were B. Harrison (Captain), W. Burr, J. Harrison, E. Weaver, and T. Wright) in competition with the Ambulance Corps of the Normanton District, viz.:—St. John's Colliery, Whitwood Colliery, Normanton Railway Station, and Sandal Railway Station.

First Aid was encouraged; each man who qualified received a certificate.

'Gentlemen, most of you have seen the WEG. Take what you like that is good, and throw away the bad. Drop the name, and let us have a British apparatus,'

Shortly afterwards he was once again invited to give evidence to another Royal Commission: this time the subject was coal supplies.

One day, as he walked through the village he overheard the following conversation:

'Why did you have to tear up your best working shirt,' Harriet Burr's angry voice demanded of Bill.

'I had to stop a mate's bleeding.' Bill answered.

The following afternoon, Bill's daughter Lucy and some friends played with skipping ropes and spinning tops as they waited at the colliery gate for their fathers. Bill's coal blackened hand had just been taken hold of by Lucy when Mr Garforth approached.

'Can I have a word with you Bill,' Mr Garforth called. Bill left Lucy momentarily, 'I hope that will cover the shirt,' he gave Bill sufficient money for two shirts, then added, 'I have arranged that in future all first aider's will carry a small pouch containing bandages.'

In 1903, colliery workshops were experimenting with flameproof motors; they filled the motor-chamber with gas and ran them for 6 hours.

The summer presentation of first-aid certificates and trained Mines Rescuer certificates for 1904 was held at the Garforth's new home, Snydale Hall on the outskirts of Normanton.

Hearing the cries of delight from the local scout troop who had camped overnight, Mary Garforth looked through the window to see that preparations were under way. All her six daughters were there. The youngest ones, Elizabeth and Stephanie, enjoyed the company of people of their own age. Margaret and Helen, the eldest, supervised the servants, while Frances and Christina helped.

Men trained on the colliery surface using equipment part-sponsored by the Garforths.

To improve their skills, competitions involving many teams were organised.

Snydale Hall, now the home of the expanding Garforth family.

She looked across at William, who had been collating his notes after once again receiving a summons from another Royal Commission, this time on the use of electricity underground.

'Everything is getting so expensive,' Mary said as she resumed her checking of the household accounts, 'Milk 3*d* for two pints, candles 3*d* per dozen, butter 9*d* per lb, 6*d* for a dozen eggs and clogs 1*s* 11*d* per pair.' William was listening, so she continued, 'However can the villages afford to buy calico to use for making underclothes at 4*d* per yard. What's this, children's shoes for 2*s* 11*d* a pair? That's why so many are bare footed and the pawn shops are always busy.' That year the village was invited to a garden party at Snydale Hall, just three miles from Altofts

'Gentlemen,' William, several weeks later was telling an audience at a mining meeting in Manchester, 'we think we have solved the problem of a cage crashing down the shaft,' and he went on to explain. 'We have mounted two bell-shaped housings on the cage headgear and between the rope and the cages we have attached another unit. Now, if control of the cage is lost, the weight of one cage crashing down the shaft will force the unit on the other to fly into the bell-shaped housing, shearing off a copper pin on this unit. This would release the cage rope and two metal arms in the unit would spring out and clamp the cage in the head gear. Also, we have built into the head-gear a series of metal blocks which allow the cage to go up but not down. Full technical details will of course be available to you.'

Also he revealed that the rescue teams at Altofts using the new WEG and the new German Meyer & Giersberg rescue apparatus were now able to work for two hours in carbon dioxide filled galleries. Finally, he told them, 'If you ask us to help you, we will come.'

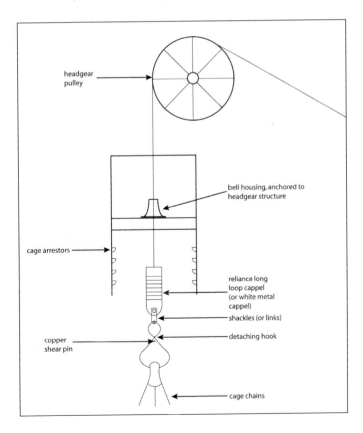

eleven

1905-1907
THE RESCUE TEAMS ON ALERT

The daily newspaper headlines read in early March 1905:

OVER 1100 KILLED IN MINING DISASTER AT THE COURIER MINE IN FRANCE

Mr G A Meyer, one of the foremost German mining engineers of the day, and a designer of mines rescue apparatus, said today, if Mr William Garforth's rules for the recovering of a coal mine after an explosion had been used, lives would have been saved.

The editor of this newspaper commented:

These rules devised by Mr Garforth have been distributed at his own expense to every colliery in Great Britain.

Mr Meyer then went on to say:

Men lacked confidence in the rescue apparatus they were wearing and actually took them off, causing more deaths. If only mine rescue training galleries like the one Mr Garforth has erected for his employees, could be erected all over the world, no doubt many more lives would be saved.

The editor went on to say:

The world's first training gallery was set up at the West Riding Colliery at Altofts, Yorkshire, the building and training of the Mine Rescue team financed by Mr Garforth. Since the training centre opened, mining engineers from all over the world have visited and been impressed by the work being undertaken there, by both management and men. All of these men give up their free time in the hope of saving their fellow men's lives.

The summer of 1905 was a happy time for the village of Altofts. The rescue teams at the West Riding Colliery, unaware of the comments, carried on training and improving the various rescue apparatus' and their skills.

The men were working five days a week and every penny was vital to the Burr family as they struggled to meet the mounting costs as Harriet gave birth to their fifth daughter, Dora. At Snydale Hall, a butler, ladies and kitchen maids, and three gardeners, all looked after the needs of the Garforth family.

Mr Garforth was still experimenting with portable electric light underground as more and more miners were suffering with nystagmus, the poor light underground making their eyes start to oscillate. But they could not overcome the problem of ensuring a spark did not cause an explosion.

The Wakefield factory had received so many orders that Mr Garforth employed more men and considered extending the factory. Around the same time, Helen Garforth announce her engagement.

Now we move to Parliament.

'We must nationalise the mines.' stated one of the sixteen new mining MPs, after the latest elections to the house. 'Over 1 million miners are working only one or two days a week as the industry goes through yet another depression.'

Once again, depression brought hardship to the area. Families queued for hours, shivering in the snow to collect a small parcel of food from the colliery school. They used the last of their savings to make a good meal for their children on Christmas Day.

Once again every member of the Garforth household was involved in helping the needy.

'Tomorrow will be New Years Eve!' said William on his sixtieth birthday, as the thick soup he ladled splashed into the bowls held up by hungry youngsters. Elizabeth handed slices of bread to small, grubby hands, while Stephanie, his youngest daughter, poured warm drinks. William looked as he heard one small boy licking his bowl clean, his eyes travelling along those who queued patiently. His gaze reached the open door as, another girl shivering under a snow covered coat moved into the colliery school room. 'How many more outside?' he thought, 'Is this the only warm place in the village?'

'Let us hope that 1906 brings happier times,' Mary said as she joined a group of desperate mothers who were waiting outside, ready to take their children back to cold homes. She shivered as she felt the warmth of the school room on her return, then stamping the snow off her boots she moved over to William. 'Let us hope no more epidemics break out.' She brushed the snow off her coat, 'Have you managed to persuade the local boards to combine and build a hospital where new cases can be isolated?'

'How do children get so badly burnt?' Elizabeth asked before he could reply, the vision of a youngster they had served troubling her. William looked along the line of shuffling, shivering children, then, raising his voice, 'Young children are not aware of any danger, they stand too close to the fire and get burnt.' As the last of the children lifted his bowl towards William, he said, 'Heat is a blessing but it can also cause all kinds of problems.'

'Mr Garforth,' a call came from the door.

'What did you mean, papa, when you said heat caused all kinds of problems?' asked Elizabeth as she pulled a blanket around her in the carriage on the way home.

'Our New Moss Colliery at Ashton-Under-Lyne is, as you know, the deepest mine in Great Britain. In places the temperature is well over ninety degrees Fahrenheit.' he answered.

'My goodness!' said Stephanie as she placed her bootless feet on a hot water bottle.

'I have just ordered a cooling plant so that we can reduce the temperature to seventy-five degrees. It is the first one that has been designed for mines.'

'Papa, if this is the first one, how do you know what to do?' asked Frances. William answered,

The wedding of Helen Garforth was not forgotten by local people for many years.

'The air going into these places will be directed by means of canvas screens through the cooling plant.'

Mr Garforth was asked by the government to be Chairman of a standing Committee on Scientific and Industrial Research.

'OVER 1 MILLION MEN WORK IN THE NATION'S COAL MINES, ONE IN EVERY 886 WILL BE KILLED' the newspaper headlines read.

'When is the government going to do something?' demanded Mr Fred Hall, Normanton's MP, brandishing his newspaper in Parliament, as sixteen mining MP's shouted their support.

'We have arranged for a Royal Commission on Accidents in Coal Mines,' answered a spokesman for the government, 'they will begin taking evidence in a few weeks time.'

Mr Garforth explained to the Royal Commission that his collieries were amongst the safest in Britain because they had introduced and trained men in many new safety features. He gave examples of how lives had been saved.

'We had two underground explosions, but because we had trained men, no one was injured.' He went on in great detail to explain the many safety features he had introduced, and concluded by saying, 'I believe an educated man can fully understand the dangers; that is why I encourage all my employees to attend night school.'

Mr Garforth was also now Chairman of the Governors of the newly built Normanton Grammar School, and was trying to persuade the local councillors to build a girls' high school.

In 1906 the Mining Association of Great Britain announced that the Scottish area had been testing Mines Rescue Apparatus. Their conclusion was that, 'The WEG is the best suited for use in our mines, being lighter, smaller and more flexible than any others'. They also proposed that galleries should be built to represent a reasonable length of underground roadway, and that experiments be carried out to prove or disprove Mr Garforth's theory. They also proposed that he should be entrusted with the responsibility of carrying out experiments on coal-dust.

In 1907, Mr William Edward Garforth was elected the President of the British Mining Association. His first announcement was that His Majesty, King Edward, had created a new medal for gallantry in Mines and Quarries. The warrant stated that the king was desirous of distinguishing by some mark of royal favour heroic acts in the mines and quarries industry.

Normanton's new library was built, with a very large donation coming from Mr Garforth. On opening it, Mr Fred Hall, Normanton's MP, said, 'Mr Garforth and I are opposed to each other in some matters. However, I want it clearly understood that there is no man I admire more. He takes a deep interest in everything that has for its object the bettering of conditions for the working people.'

Helen Garforth.

Mr Garforth was also asked to accept the Chairmanship of a Committee of Magistrates, appointed to look into the possibility of reforming young offenders.

Meanwhile, the Mines Rescue training followed his familiar routine.

On 1 October 1907, the Royal Commission on rescue apparatus in mines gave their report, which read,

'Of all the apparatus tested, only two could be marked as good; The Fleus-Siebe Gorman, renamed the Proto, and the WEG.'

On the morning of 15 October, the telephone rang at the Altofts and Tankersly Rescue Stations; a fire had broken out underground at the Warncliffe Silkstone Colliery. While a wagon was loaded, a short sharp blast, the prearranged signal, was sounded on the colliery buzzer. Wives and mothers rushed upstairs; the standby team had heard and work trousers were already being pulled on. Sandwiches were snatched as they clattered down wooden stairs and across stone-flagged floors, and within minutes they had run to the colliery. They got onto the first available train with their apparatus. Almost before the steam had started swirling around them at their destination, the apparatus had been unloaded and placed onto waiting horse-drawn wagons. The horses seem to sense the urgency, and with angry snorts, they clattered over the cobbled streets and out into a muddy country lane.

However, the journey by the West Riding Colliery team was not needed. The nearby Tankersly Rescue Team had arrived at the colliery in forty-five minutes and had the fire under control before the Altofts team arrived.

In December, Helen Garforth married. Families cheered the mile from Snydale Hall to Normanton. Crowds lined the road from Normanton to the Altofts village church. For thirty minutes, a procession of previously shiny carriage wheels, now covered in mud, passed through the excited crowds. Every window overlooking the route was full. Each carriage in turn allowed its passengers to alight onto the carpet of straw and flowers that had been laid under the tunnel of white flowers at the entrance to Altofts Parish church. Gentlemen in morning suits removed their top hats as they and their ladies, in very fashionable dresses, stirred up the carpet of flowers lining the floor into the flower decorated church. In a large carriage open to the elements, Mary and her four youngest daughters acknowledged the cheers. The eldest two daughters and four friends were bridesmaids, and arrived in a similar carriage just behind.

Their dresses were white crepe-de-chine over pale pink silk, and they were wearing white satin hats. Each was carrying a bouquet of pale pink carnations. The crowds buzzed excitedly as the minutes ticked by. Helen allowed the people to share her day, taking advantage of the winter sun; her carriage roof was folded back. She wore a white satin dress with a yoke of Brussels lace trimmed around the hem and up the front with silver embroidery. William, wearing a dark morning suit and top hat, sat quietly beside her, enjoying the moment.

That evening, the band stopped as all heads turned towards the door. Mr Garforth, his daughter Helen and her new husband entered the fairy land that had been created in the colliery school-rooms. He had provided beer for the celebration and Helen moved from table to table, introducing her husband. Once all had been introduced, they left to allow their employees and families to celebrate.

twelve

AN EXPLOSIVE 1908

The industry was having a series of cage accidents as cage-ropes broke under the strain and plunged back down the shaft.

'Afternoon,' each man said, the first of their safety tokens dropped into the banksman's outstretched hand as they entered the cage, then, transferring their glowing oil lamps to their broad leather belts, each turned to face the cage entrance.

'No home to go to John?' someone asked as the cage gate clanged shut, he pushed backwards, the oil lamp of the person behind warming his backside.

'Rescue training,' Answered John Welsby, the sweat from his recent labours on the coal face had left white channels on his black face. Now his body cooled rapidly as the cool air being drawn down the shaft flowed over him. Then with a clank and a rumble, the cage descended. John sat cross-legged in the shelter of the shaft headgear, coal-dust from his hands adding an extra layer to his sandwich. The rumble of the cage died away, then it began to grow.

'Afternoon,' he said over the clinks of the safety tokens which dropped into his hand. He ate the last of his sandwich, and joined the group of tired dirty men in the sun of the colliery yard.

At the Hampstead Colliery, near Birmingham, another afternoon shift of over thirty men was also going down their mine. The main shaft was out of action due to repairs, so the men were travelling down in the small shaft, six at a time. The first six men down disturbed a box of candles placed near a wooded support. A spark from the steel on one of the clogs of the second six caused a candle to ignite and the third team were held up slightly at the surface. When they descended, roaring flames greeted them.

'I'll warn the men who have gone int' work,' said one and set off, as a burning wooden support groaned and cracked.

'Get them fire buckets!' Dust-skimmed water passed down a line.

'Signal the cage back down!' The bell rang. There was a clang as the cage came to rest at the pit bottom. 'Leave it!'

The sweating team, crammed into the cage, made way for another. He felt the heat as the man reached out and pressed the signal button.

'Manager wants a report,' said the banksman as they re-emerged.

'We have a fire underground near the small shaft. Repairs in the main shaft must be stopped and the cage re-slung,' the manager said over the telephone, after they had reported.

Meanwhile at Altofts, John enjoyed the sun as he walked over to the Mines Rescue training school.

'Afternoon John,' other members of the team called as they arrived.

'Afternoon, James. Forty-one,' John grunted, 'Forty-two,' as he carried on lifting weights. James Whittingham, James Cranswick, Charles Haslop, and James Hopwood joined in the weight-training routine.

Hamstead Miners message.

'Fire!' The lone miner called out at Hampstead, as he caught up with the team who had entered the mine just before them, and with chest heaving he explained, and they set off back. The smoke increased; water bottles were poured over sweat cloths and held over mouths and noses. Wood cracked, creaked and burned; the remainder of the water flowed over each man, then they dashed through the flames to the cage. Ears popped, due to the difference in pressure as the cage rose through the flowing ventilation to the surface. That same flow of ventilation allowed the deadly carbon monoxide produced by the fire to snake its way through the underground roadways towards the twenty-plus men still left in the mine.

The Altofts team helped each other don the leather shoulder smocks, and then checked each others' breathing apparatus before going into the smoke filled training gallery. Two hours later, sweating fingers held a pencil to fill in the report book.

'Home time,' John said as he shivered in the cold night air, then took his old overcoat from where he had hung it and covered his grimy shoulders.

'See you next week,' each one called into the wind, as they separated from the walking group to turn into their own dark, muddy street.

At the Hampstead Colliery the main shaft cage was ready. 'Get a blackboard,' the manager told his surface foreman. 'Write on it,' the manager handed over a piece of chalk, 'signal if you are there.' He signalled on the bell push and was answered by the winding engineman's bell signal back. The steam engine took up the tension on the cage rope; pulley wheels rattled as it descended then stopped. All stood quietly. Minutes passed, then half an hour. No signal.

'Signal the cage back up, put a linnet on and send it back down. Give it ten minutes, then pull it back up.' Once again they obeyed the manager's instructions.

'The linnet came back up alive,' someone reported to the manager.

'Right, keep the cage at the shaft bottom,' the manager said looking up from the plans of the underground workings he had been studying, 'Every fifteen minutes, pull it up and send a linnet or rat down to test the air.'

At his home in Greenbank Road, Altofts, John Welsby called to his wife Harriet, 'I've a headache, can't you keep the kids quiet?' Nellie, aged twelve, and eight year old Cyril played excitedly with friends in the front room as Harriet did not want the mud from the street carried to her spotless floor.

'They're only playing,' she answered. Last week, she had been so worried by the headaches John suffered after he had been training, she had asked Jack Dyson why he suffered from them.

'It's the build up of carbon monoxide in the body from being in a gassy atmosphere. Once out in the fresh air, the body slowly clears it out of the system,' Jack had explained. As the last chime of ten died away, John placed his arms around pregnant Harriet.

'I love you,' he kissed her, felt the moving unborn child, then climbed the stairs to bed.

All through the night the Hampstead cage dropped into the darkness and each time the rat or linnet came back alive, but no men.

John Welsby had a good night's sleep, and set off for work around 5.00 a.m. By 7 o' clock, working on his knees in the light of his oil lamp, he had completed the filling of the first of his coal tubs.

The WEG rescue team from the West Riding Colliery.

'Jack, there is a serious fire at Hampstead Colliery, near Birmingham. Get everything ready,' Mr Garforth shouted from his horse and trap. By the end of the shift all was ready.

John, unaware of the disaster that had occurred, picked up the carbolic soap, and stepped into the tin bath placed in front of his coal fire.

'Another shift over,' he thought as he felt the warmth.

'If only John would wash his back and stop the bedding getting so dirty,' Harriet thought as she hung out the newly washed sheets.

The howl of the colliery buzzer shattered the silence, her heart stopped. For a second, Altofts froze. All knew what that signal meant; a mining disaster had occurred and their men had been asked to help.

'It's our team that's on standby,' John said excitedly as she entered the kitchen, the fire spitting as water splashed on to it as he stood up. Harriet, unable to speak, gave him a towel, composed herself and placed his dinner on the well-scrubbed kitchen table.

At the West Riding Colliery, a wagon and its horse were already harnessed and loaded up. 'Jack, take some money for the train fare, round up the team on standby and report to the station, along with the equipment for the Birmingham train.'

'Yes Mr Garforth,' answered Jack. He had already sent word to Dorothy.

'All here, good,' Jack ticked off the names of the Rescue Team, John Welsby, Charles Haslop, James Whittingham, James Hopwood and James Cranswick. He then checked with Thomas Fox that all the apparatus and other equipment were loaded safely on to the train. This was an exciting journey into the unknown for them all. They travelled the last few miles in cars that had been waiting for them.

'Make way for the rescue teams,' the driver called as they neared the colliery, the dense anxious crowds divided to let them through.

'Thank god,' said one young wife whose husband was underground. A small number of people cheered but most stood silent, their eyes following the team.

'Meet the Tankersly Mines Rescue team,' the Hampstead Colliery manager said as they unloaded their apparatus.

'Only two of you?' Jack asked.

Against all their team rules, they made the decision to split the Altofts team.

'These are the roadways where the men should be,' the manager informed them as they peered at plans. They finalised their route, then, donning their Mines Rescue apparatus, both teams moved to the cage. Sweat was soon running as the teams moved their separate ways along the partially collapsed underground roadways. Still-burning wooden supports added another danger. Weakened, some had snapped under the strain and the whispering rocks had filled the gap. They cleared a way, supported the roof, then squeezed through. Once clear, they checked each other's apparatus for any damage. After one hour of searching, as per their training, they started their return journey.

'What's conditions like?' Jack asked. Two hours had seemed like an eternity as he waited near the cage entrance.

'We are having to walk at a crouch, or crawl, and our apparatus is constantly catching on jagged pieces sticking out from the roof. Also there is a lot of rubble lying around, so we are having to watch our feet,' James Whittingham answered as Thomas Fox refilled their oxygen cylinders.

'It's just like home,' John added.

'Have you found anybody?' anxious, gaunt-faced relatives asked from the watching crowd.

Relatives waiting for news after
the explosion at Hamstead
Colliery near Birmingham.

'Have a longer rest, lads,' Jack urged, but the team were already donning their equipment. Down they went again, as the clock at the cage-side ticked on. Two more hours.

'Anything?' the cage banksman asked as they surfaced again.

'No.' answered James Hopwood, pulling the breathing unit away from his mouth to breathe in the cool fresh air. Jack and Thomas moved in quickly to help them off with their apparatus as they sank to the floor. Pots of tea and sandwiches were brought to them and exhausted, one by one, they laid back and went to sleep. Jack and Thomas filled their cylinders again.

Meanwhile the thousands-strong crowd all around the rescue teams shuffled and stared, waiting for news. These exhausted men were their husbands', sons', or friends' only chance. Thirty minutes passed; the relatives surrounding the rescuers constantly changed, but they asked the same question:

'Any news?'

'One more try,' said James Hopwood as he stood up. A sigh went through the crowd.

'You must have more rest, lads. What about the build up of carbon monoxide in you?' Jack pointed out

'I feel fine,' said James Cranswick as he looked at the anxious faces surrounding them. The crowd watched in silence as the cage wheels rumbled their descent.

Once underground, weary limbs set off again. John Welsby and James Whittingham again consulted their map. The cross road at which they stood seemed to be getting deeper in water as they made their decision.

'That way,' James signalled, pointing to a road that was several inches under water. The road began to slope up and the fast-running water made the floor very slippery. The damp floor had stopped the passage of the fire ball, so the roof was secure, but slowly the mud took its toll on their diminishing energy. Each time one slipped, only will power and training told tired limbs to get up, until finally John could not muster the strength and reached down and pressed his horn. James stopped, rubbed the mud from his eyes, and chest heaving, he held a wooden support for a few seconds, then turned round and started back towards John. Dropping to his knees in the mud, he checked John's cylinders and

PUBLISHED BY W. GOTHARD, 6, ELDON STREET, BARNSLEY.

THE YORKSHIRE HERO.

JOHN WELSBY

DIED MARCH 5TH 1908   AGED 31

WHILE ENDEAVOURING TO RESCUE ENTOMBED MINERS AT THE HAMSTEAD COLLIERY LOST HIS OWN LIFE

John Welsby, a member of the team, killed whilst trying to rescue trapped men underground.

pipes, then somehow together they staggered up and on. Stumbling several times, finally James could no longer lift John.

'Must get help' his brain said as he allowed his instinct to move him forward, eventually collapsing at a cross road.

Meanwhile, as the other teams were also making their way back towards the pit bottom, one almost fell over James. While one checked him for signs of life, the other searched along the other roadways for John. Then, glancing at their watch, they pulled one of James' arms over their shoulder and carried him to the cage.

Jack watched the minutes tick by, his heart stopping as he heard the cage signal-bell. The answering winding engineman-bell, and the rumble of the pulley wheels, silenced the crowd. Every eye counted the men off the cage; there was an intake of breath as Jack and Thomas moved in and lifted up a body that had been hung over the side. Others moved in and removed their apparatus as the rescuers sank to the floor.

'We found him laid out exhausted and carried him to the cage,' one gasped as Jack removed the rescue apparatus from the unconscious James Whittingham. In the silence, a cylinder fell to the floor.

'He's breathing.' Jack said, rising up from his knees and covered James with an offered blanket, then pulled out his pocket watch. All eyes looked towards him. John Welsby was still underground and his life-saving cylinders had run out.

Automatically Jack and Thomas once again filled up the cylinders. The teams had collapsed into exhausted sleep.

Jack and Thomas carefully checked all the apparatus, then sat quietly, the noise of the crowd building up as the time passed.

'There's one missing from the team lads,' James Cranswick said, as he staggered to his feet. Without a word Jack moved in to help as they again donned the apparatus, and moved into the cage.

Eventually James Whittingham opened his eyes. 'Where's rest o' team?' he asked.

'Gone back for John,' Jack replied, stopping James as he tried to pull on his apparatus. No word was spoken for a long period.

'What happened?' Jack asked eventually.

'It was really hard going. Then I heard John's horn, and looked around and saw him trying to get up out of the mud. I helped him for a while,' James paused, 'but then I had to leave him. I don't remember anything after that.' The silence was resumed.

'Stay there,' Jack ordered James, as he attempted to get up on hearing the cage-bell. Pulley-wheels once again turned and stopped. Helpers moved in as they staggered off the cage. No John Welsby.

'Any chance of getting to him?' the manager asked as he moved over. His eyes followed Jack's as he looked at the men lay on the cage-side floor.

'Seal off the mine.' The decision was made.

Jack reported to Mr Garforth immediately on his arrival back at Altofts.

'William, what's happened?' Mary asked when William returned from his doorstep conversation.

'They went against all the rules and split the team.'

'What has happened?' Mary asked again.

'As a team of five, they are strong enough to work in relays of two to help an injured team member out of the mine, but because only two from Tankersly Mines Rescue arrived, a decision was made to split the team.' Mary waited, 'John Welsby is dead.'

'I will go to his Harriet,' Mary said, putting on a black coat over her dark brown dress.

On 20 March, as the hissing steam momentarily cleared, members of the rescue team moved towards the train which had just arrived at Normanton Station. When the door of the guards' van banged open, they moved in.

'Up!' John Welsby's coffin was hoisted to their shoulders and momentarily, everyone froze. Curious, silent passengers stood and looked on as the broad shouldered men, in their Sunday suits and polished boots, carried the plain coffin to the lift, then out to the waiting carriage. All the way to his home, men removed their caps, and all talking ceased as the escorted carriage passed by.

'Greater love hath no man than this, that a man lay down his life for his friends.' Mary Garforth read the last of the telegrams to Harriet and the children, including ones from the Queen, the Prince of Wales and the Mayor of Birmingham.

'He died of a noble heart,' Harriet said, tears bubbling from her eyes as she placed her arms around the children, then looked down at her swollen belly, 'I'm sure my new child will be a boy. His name will be John.'

'William will make sure you want for nothing,' said Mary looking at the group, knowing a fund had already been started and they had made a large donation.

The following morning, 250 men of the West Riding Colliery, all wearing black and white rosettes, preceded the hearse, followed by members of the colliery and Normanton Station ambulance teams. Four large carriages carried the family and the lady mourners.

The coffin was carried from Normanton Station to his home by fellow workers. Here it starts on its way to the Altofts church.

The proceeds from these postcards and other items were donated to the fund raised for his family.

Note the mass crowds.

Mr Garforth leads the Cortege.

Over 50,000 people lined the route.

Altofts church. Thousands waited patiently outside..

The landscape outside the church was touched up in this postcard, so as not to distract from the funeral.

IN MEMORY OF "WELSBY" THE ALTOFTS HERO 1908

The grave.

Mr Garforth, along with his Colliery officials and Normanton's MP Mr Fred Hall, members of the local boards, representatives from the Miners Association, and friends of John and Harriet from Liversedge, walked with the coffin. Almost fifty thousand people lined the short distance to the Altofts church and cemetery. Harriet was provided with a chair while the many wreaths were laid at the grave after the service.

The newspaper headlines read:

THE BODY OF THE HERO JOHN WELSBY WAS FOUND ONLY YARDS FROM WHERE RESCUERS TURNED BACK

It went on to state:

At the inquest into the death of the twenty-four men at Hampstead Colliery, it was discovered that only four of the dead men had known of the danger. As they began feeling the effects of the colourless, odourless, and tasteless carbon monoxide, they had managed to stagger only a few yards from their working place. The rest died as they worked.

William Garforth stated,

'One visit in those small, confined roadways wearing rescue apparatus would have completely exhausted most men. To complete three or four just asked too much.'

The newspapers added:

'If only there had been more rescue teams, these men would not have felt they had to keep trying.'

The men of the West Riding Colliery Mines Rescue teams carried on their voluntary training with the same determination. Mr Garforth received a letter inviting him and the men involved in the Hampstead Colliery rescue to Buckingham Palace on the 23 July, in order to receive the Edward VIII Medal for Conspicuous Bravery from His Majesty.

On the 11th of July Mr William Garforth, as Chairman of the Leeds University Advisory Committee, was presented to the Prince of Wales.

In Birmingham, at almost the same moment, the Mayor was addressing a large crowd, 'My Lords, ladies and gentlemen, it is my very great honour to award these gold and

The monument raised by public subscription.

King Edward presented his medal (The Miners VC) to the team, and Mrs Welsby, at Buckingham Palace.

A napkin, now over 100 years old.

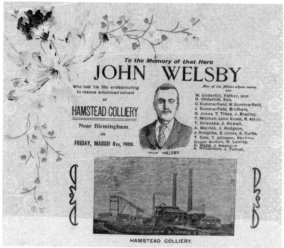

These napkins (in the author's collection) were sold to raise money for the widow.

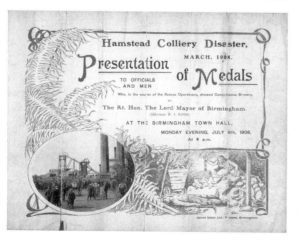

The Mayor of Birmingham presented the rescue team and Mrs Welsby with a certificate.

silver medals to the heroes of the Hampstead Colliery disaster.' The men had one by one advanced to the cheers of the crowd. Then Harriet Welsby, to greater cheers, was beckoned forward. The Mayor held up his hand, and as the cheering subsided and she sat down, he continued, 'We have, besides making a donation to the Colliery Disaster Fund, also made a donation towards the fund which stands at £1,000 to help Mrs Welsby.' Harriet grasped her children's hands and they all stood up.

'Jack, have you heard?' asked Bill Burr, 'When the lads went to London, Fred Hall arranged for them to stay at a right posh hotel, and next day they went in a car to Buckingham Palace.' Jack waited. The news had quickly passed through the village, but he let Bill tell him again. 'It was the King's birthday when they went for their medals, and the place was full of toffs. They told me they stood caps in hands, until a voice called each one of them by name. They then had to walk forward, bow, and receive their medals, and everybody applauded them.'

'Aye,' said Jack, 'Harriet Welsby has been telling everyone the Queen spoke to her as she received one for her husband. She was so proud.'

As a series of short sharp blasts sounded on the colliery buzzer, men stopped all access to lower Altofts.

'Everybody out!' Fists thumped on doors as the message was shouted throughout the West Riding Colliery village. Jack stood up, water splashing from his naked body to join the black water that swirled around at his feet in the tin bath.

'Everybody out!' As the banging on the doors moved gradually down the street to his, Jack's feet left wet footprints on the stone-flagged kitchen floor as he reached for a towel.

'Everybody out!' Harriet Burr transferred her bread from the fire-side oven to the kitchen table.

'Everybody out!' As the shouts disappeared, mothers carrying crying babies, fathers tightly clasping young hands, and older children, some running, others helping their grandparents along the uneven paths, poured into the streets and made their way into the open countryside beyond the village.

Young Robert Beecher and friends lay low in the tall grass on the railway embankment as a colliery official searched the area. Then they peered out at all the activity in the colliery yard as they heard a strange rumble. Suddenly a red flame shot out of the end of a metal tunnel of the old Lancashire boilers that had been welded together on the colliery surface. They clamped their hands to their ears as the blast passed them, and as they stared, large jagged objects floated over their heads like pieces of paper.

Like a gigantic pressure hose, the blast moved through the village, pushing dust and other objects before it. Windows were shattered, newly-baked bread sank. As the blast reached some horses pulling a self-bidding machine in a nearby cornfield, they bolted, trampling over a farm labourer.

Robert and his friends climbed down the railway embankment and ran to where one piece had gently landed, 200 yards away from the galleries. 'Oh my...' he trailed off, looking at a large piece of jagged edged metal as it smouldered in the grass. The lads scampered back to their vantage point just as the men in Mines Rescue apparatus, carrying canaries, moved into the badly damaged structure.

Later, they watched as the many spectators at the demonstration moved off in their carriages and cars; it was an exciting time in the village.

The next day, the papers reported:

Team from Pope & Pearson, Ltd., Rescue Station.

W. T. WEBSTER   W. BURR   S. BERRY, Captain   U. HOLLAND   J. WHITTINGHAM
            J. HOPWOOD, Engineer   SERGT. BYRNE, Instructor   (King Edward Medal)
                (King Edward Medal)

The teams were available to help
at other collieries as explosions
continued to happen. All volunteers'
expenses were paid by the
Garforths.

'EXPLOSION HEARD SEVEN MILES AWAY'.

One went on to explain:

> The Mining Federation of Great Britain has raised £10,000 from its members, for experiments
> which are to be carried out at The West Riding Colliery at Altofts. The objectives are:
> 1) To demonstrate the explosive nature of coal-dust and air, without the presence of
> inflammable gas.
> 2) The discovery of a preventative.
>   Mr William Garforth has been given the honour of supervising these experiments. Mining
> engineers from Germany, France and America have already visited.

The item went on to give a full explanation of what was happening.

Young Robert and his friends watched as, over the next few days, more Lancashire
boilers were cut and then welded to replace the damaged ones. Once the 1000-feet long
metal roadway was completed, it was strengthened by iron hoops, and then continuous
laps of chain were wrapped around it. Then once again wooden shelves were fixed to its
walls and filled with coal-dust.

'Gentlemen, the coal-dust you see around is from the Silkstone Seam. However, we
have proved that dust from Scotland or even South Africa reacts in the same way,' Mr
Garforth announced as he once again conducted a party through the metal tunnel.
'The dust is spread evenly along the floor and walls as if it had dropped from coal tubs.'
As they walked along he added, 'The ventilation is supplied by an exhaust fan that
produces 80,000 cubic ft per minute.' He looked at his watch as the short, sharp blasts
began.

'Could you all please retreat beyond the barrier.' He escorted the spectators several
hundred yards from the metal tunnels, as his men escorted the villagers to safety. 'You will
hear two shots from the small cannons you have seen,' William waited as he once again

Other members of the team were presented to visiting Royalty.

had their attention, 'both to simulate shots fired underground. Because of the slight delay between them, the second one will be fired into a coal-dust filled atmosphere.'

Two loud bangs, followed by a rumble of sound, moved down the vibrating metal tunnels until an enormous red flash of flame shot many yards out of the end; the following blast pressed them back into their seats. Then, through the smoke, almost in front of the astonished spectators, large pieces of metal gently floated to the ground.

'We are using every opportunity to give our Mines Rescue team the chance to train under the conditions they will experience underground,' William said as the teams moved into the tunnel, then he allowed the spectators, ears still ringing, to move forward.

'What force it must need to blast such a piece of metal from the side of the tunnel and transport it here,' one said, as they stopped at a large piece almost 100 yards from the tunnel.

'Now I believe coal-dust can cause explosions,' another astonished spectator said. Meanwhile, flag-waving men moved through the village shouting the all clear.

William reported back to the British Mining Federation, 'Gentlemen, most of you have visited the experiments we have been carrying out at Altofts. I believe we now have a preventative. We mix stone-dust with the coal-dust.' After a short pause as this information was digested, William added, 'May I have your permission to carry out the second part of our objectives?'

The experiments were carried on throughout the summer. However, this time stone-dust was spread around and no explosions occurred.

On 1 August, the West Riding Mines Rescue heroes, along with some trick cyclists, led the parade that opened the village carnival.

Two weeks later the newspapers reported:
'The British coalfields are again covered by an extensive area of high barometric pressure. It is highly desirable that persons employed underground should pay close attention to any indication that inflammable gas is leaking from the strata.'

By 18 August, the difference in pressure was allowing gas to seep out of old workings into the roadways at many collieries. At the Maypole Colliery in Wigan, a spark ignited

In 1908, £10,000 was raised (equal to many millions now) and Mr Garforth was appointed to carry out experiments to prove coal-dust could cause underground explosions. Old Lancashire boilers had their ends cut off, and welded together.

The coal-dust was placed on shelves that ran the length of the galleries, to simulate an underground roadway, and various items placed along the roadway.

The ends of the galleries were sealed.

Cannons were fired to simulate an underground explosion.

Some explosions could be heard seven miles away.

Debris was scattered over a wide area.

As each new exhibition was set up, repairs were carried out. Then the total population of the colliery village was asked to leave.

The cannons fired.

The mine owners and management who came from all over Europe now were convinced.

the gas. The vibration from the explosion could be felt on the surface almost a mile away. Within minutes, the colliery yard was crowded with workers' relatives, clambering for information. Fortunately, the hundreds on the day shift had already left the mine, but seventy-eight men were still underground.

A volunteer-rescue party quickly moved into the mine and managed to rescue three men.

'Gas!' An official shouted, looking at his oil lamp. With no breathing apparatus, they were forced to leave the mine. Once out, they anxiously monitored the outgoing air for smoke, very relieved that none came.

'It should have cleared now,' an official announced after allowing time for the flow of clean air to clear the workings.

'Look at this,' the first in the rescue party said, as he looked around at the badly burnt bodies. One turned away and was sick.

'He has got his hands in front of his face, they must have seen the flames coming.' another commented, and after checking for gas again, they moved on. After another 100 yards, they found two who had been praying.

By late afternoon the following day, seven bodies had been brought out, and more were being recovered, when all around them began vibrating. Rocks of various sizes thudded to the floor; slices of the wall slid down; wooden supports creaked, groaned and some split. The planks they had held up dropped at an angle, followed slowly by an area of the roof. More rocks dropped as the vibrations continued.

'There's been another explosion!' someone shouted, and they ran for their lives. Eventually, chests heaving, all forty-two men crammed into the cage, overloading it, and signalled it up. Fortunately, it managed the load.

'There's two still down there' the banksman said as he counted their safety tokens and added, 'it's two who operate the water pump.' All stopped their movement away from the cage and looked at each other.

'We'll go for them,' one said as he looked at his friend.

'Use these.' Sweat cloths, damped from water bottles, were passed to them. As the remainder sat down to wait, the wheels started their rumble. Time passed, then the cage-bell rang.

'They were just conscious,' a voice came out of the cage as the rumbling stopped. Four men staggered from the cage. Fire was now raging in all the underground workings; the carbon monoxide it produced meant that no team without breathing apparatus could fight it.

'Send for the Mines Rescue teams.' The manager ordered. Never before had a Lancashire colliery manager been able to call out trained Mines Rescuers.

The buzzer sounded at the West Riding Colliery. The standby team were James Cranswick, James Whittingham, Charles Haslop, Samuel Berry, Abraham Whittingham, Tom Fox and Sergeant Byrne, who was also travelling to look after the equipment.

By the time they arrived, the fire had been burning for two days. Valiantly they fought the flames, but the heat had now ignited the coal. With only the Altofts men and one other team, the stranded men could not be reached. The decision was made to flood the colliery to put the fire out.

Mr Garforth was invited to Frankfurt in Germany to contribute to the first international conference on how to save lives after a colliery explosion. His talk was on his updated 'Suggested Rules for Recovering a Coal Mine after an Explosion'. In a side room, he was showing a model of the Altofts Mines Rescue School.

Certificate presented to trained Mines Rescue men.

'Ladies and gentlemen,' Mr Garforth addressed a packed audience in the colliery school-room. 'I am going to explain to you what the purpose was of the explosions we have been carrying out in the village.' With diagrams, he explained to the audience.

That Christmas, a turkey was taken by Mrs Garforth to every family who had been inconvenienced by the coal-dust experiments.

The Altofts Rescue teams were now averaging almost two training sessions a week as they strived to improve their skills and the apparatus. Mr A. Firth wrote in the Altofts Mines Rescue comments book after a training session, 'Head piece too light on nose, and when I raise my head it shoves the mouthpiece below the chin and air escapes.' Individual comments such as these were noted and efforts made to rectify the faults.

However, still many colliery managers were not convinced that coal-dust caused explosions. When explosions occurred at their colliery, they did not send for the Mines Rescue teams. There were 153 explosions underground in 1908, causing the loss of 128 lives and many hundreds of injuries.

thirteen

1909-13
AT LAST THEY TRIUMPH

1909. West Stanley Colliery, County Durham, 168 killed. Darren Colliery, Glamorganshire, twenty-seven killed.

In 1909, the Altofts Colliery football team played in their new strip purchased by Mr Garforth. With Bill Burr as captain, they won three trophies.

Sweat ran down Bill's neck as he glanced around, then resumed his stationary position. He remembered his wife's happiness when he told her he had won the draw to represent the team, and as the reality dawned, she had hugged and kissed him. It had been a happy evening, and unable to restrain her joy, the neighbours were soon sitting at her well-scrubbed table. Jugs of beer from the local public house soon had them all singing while Bill played the piano. Only the thought of getting up for work at 4a.m had brought the merriment to a halt.

Again he looked around the Entrance Hall of Leeds University as the talking died away. He felt so out of place, all these learned gentlemen in their silk top hats and formal wear. His team-mates had laughed at him in his starched shirt, pressed trousers and polished boots as they had helped him load the apparatus on the train. Once again he mentally rehearsed what he had to do. If only his parents could see him, they would be so proud. He glanced over again. The nodding top hats were deep in discussion; he shivered despite the warmth.

Silence filled the room. The top hats shuffled into line, and one by one they were removed and black coats tilted down. Then Mr Garforth and another gentleman emerged around the line.

'Your Royal Highness,' William Garforth addressed the Prince of Wales in the February of 1909, 'this is the WEG rescue apparatus. Please let me explain how it functions.' Bill bowed his head, the face mask covering his blushing face, then once again stood like a statue as Mr Garforth explained the apparatus to the Prince.

'I understand one of your teams has just returned from helping with the recovery of the Maypole Colliery after the explosion. What was it like?' the Prince asked.

'As your highness is aware, entire family's men-folk were killed by the carbon monoxide. My men found fathers who had been holding wet clothes over their son's mouths and noses, while they sat on their knee.' William stopped and looked at the Prince, then carried on. 'Others had placed wet handkerchiefs over their tea cans and tried to breathe through them.'

'Why wet?' asked the Prince.

'They believe it can stop the carbon-monoxide.' Again he looked at the Prince, 'In nearly every case, my men had been beaten to the bodies by mice.' As it dawned on the Prince what this meant, his face went white and he swallowed hard. William changed the subject.

The members of the rescue teams, family and friends all quizzed Bill when he arrived back home.

Late one evening, the following week, the alarm sounded.

'Explosion at the West Stanley Colliery in County Durham,' they were informed on their arrival at the colliery, 'Hundreds of men and boys are underground.' Once again the rescue apparatus was carefully checked and a team caught the evening train, but they were too late. Mr Garforth ensured no man lost his wages.

Many engineers wrote to Mr Garforth about their solutions to reduce the risks of coal-dust explosions. One learned engineer wrote outlining that they were using ground-up glass instead of stone-dust. Mr Garforth wrote back and explained that the mixture was insoluble in a digestive system, so would be dangerous to use in mines.

The colliery village was now used to the periodic evacuations as the coal-dust tests continued. Shattered windows were replaced as the families moved back in. The tests were suspended on the 24 August 1909 as the village buried one of their characters, William Nicholson Harris, a native of Trinidad. He had worked at Pope & Pearsons for five years before injuring himself the previous year. Tragedy again struck the village that year as a measles epidemic killed fifty. The Garforths had spent every daylight hour in the fight.

Also around this time, colliers were travelling around the country demonstrating the latest Diamond Coal-Cutting machine, which was proving to be a success.

On a beautiful May day in 1910, the rescue team enjoyed the scenery and played cards as they made their way by train to Whitehaven in Cumberland. On arriving, they moved quickly into action, unloading their apparatus. A man appeared through the billowing steam and asked loudly, 'Rescue team?' Word spread quickly and within minutes, a crowd of men gathered around them and collected their apparatus. Carefully holding each piece, they now formed a solid wall around the Rescue Team. The Altofts team looked at each other in bewilderment as one asked.

'Ready?' They were marched from the station and as they passed rows of houses, men and women stopped their conversations briefly to watch them.

'Look at that!' Bill exclaimed, looking at the Wellington Colliery headgear which stood part of the way up the side of a cliff. It was built to blend into a surrounding castle wall, making a striking relief against the sea line. As they approached, they noticed the other colliery buildings had also been built in the same style, and on the walls surrounding the castle, guns could be seen. The Altofts men stopped and looked up, causing the procession to come to a stumbling halt. Following their gaze, one said,

'That American Paul Jones spiked those guns in 1776.' Their clogs clicked on the cobbles as they crossed the colliery yard that separated the buildings from the sea. Bill asked as they stopped,

'Which way are they working the coal?' One answered,

'That way,' as he pointed at the sea. A large, strangely quiet crowd in front of them parted to let them through, but they stopped. In front of them was a group of policemen and in the silence, the leader of the procession stepped forward.

'These trained rescue men have travelled all the way from West Riding of Yorkshire. They are here to go underground and rescue our men.' A church bell rung and a large clock on a colliery building clicked forward, and then again there was silence. The tension could be felt as a man turned, looked at the speaker, then, looking straight at the men of the rescue team, said,

'Please go back home. There is nothing you can do here.'

'What has happened?' asked one of the team.

'Lord Lonsdale, the owner,' he spat, 'says no-one can still live in the underground area affected by the fire.' answered the spokesman. The police moved into a line in front of his Lordship as the spokesmen added, 'They waited thirty hours to send for a rescue team.' he said loudly. The crowd behind him shuffled menacingly forward as he lifted the rescue apparatus he had been carrying above his head,

'These men are trained to work in such conditions.' He lowered the apparatus, turned, and advanced through the crowd, the rest following him out of the colliery yard.

The rescue team were persuaded to stay on, and were the first people to be allowed into the colliery, but it was to no avail. Lord Lonsdale received a message of sympathy from the King.

Bill Burr's daughter later recounted to the author. 'I overhead my father talking about the Wellington Colliery and the rescue. He said when they broke through the wall seal that had been built to starve the colliery of oxygen, they found men sat against the wall. Some had their arms wrapped around their sons.'

On the 1 December 1910 at No. 3 Bank Pit, Hulton, Lancashire, a rock fall allowed trapped gas to leak into a pit road, and a spark caused it to ignite. It exploded along the coal-dust filled underground roadways. The vibrations could be felt on the surface miles away. Then the blast travelled up the shaft. Local people who had felt the tremors turned and looked as the sound hit them. Smoke streamed above the colliery headgear. The rescue teams recovered 344 bodies.

Mr Garforth's lecture on coal-dust explosions and the possible use of stone-dust to cure the problem, at the Fifth International Congress on Mining in Düsseldorf, Germany was well received. Now he was spending a great deal of time away from the colliery. Mr Lloyd was appointed as Colliery Manager.

In Altofts, colliery personnel continued to demonstrate their experiments to non-believers, and more windows were replaced.

In 1910, 140 explosions due to gas and coal-dust killed 1,453 men in English coal mines.

In 1911 the Archbishop of York, Dr Cosmos Lang, visited Altofts, followed a few days later by members from the German and American Institutes of Mining. As the year drew to a close, the coal-dust tests were stopped. Dr D.V Wheeler, who had been working with Mr Garforth, was appointed to carry on the tests in a somewhat quieter place, on the coast at Eskmeals, near Barrow in Furness in Cumbria

To date, it had cost them the massive sum of £15,000 to prove to the many thousands of mining engineers that coal-dust caused explosions. However, many colliery owners still could not see the need to spend money spreading stone-dust to reduce the chance of explosions. A possible reason was that they had never had an explosion.

In 1911, William Garforth was elected the President of the National Institute of Mining Engineers. At that time, approximately 1.5 million men worked in the collieries of Great Britain which were controlled by these engineers.

As the industry once again went through a depression, the Lloyd family helped the Garforth women to supply breakfast to starving schoolchildren. Mrs Lloyd was formerly Miss Wordsworth of Hartley Hall in Worcester. Miss Pelia Dola was selected at the school by Mrs Lloyd, and worked for many years for the family as a cook and maid. She described

Mrs Lloyd as 'a very beautiful, tall, fair lady.' When Pelia married, Mrs Lloyd took her to Marshall & Snelgroves in Leeds and bought her a wedding dress.

In November 1911, the Miners Federation of Great Britain called a conference. The conference voted against the principal of a minimum wage. The miners consequently organised a national vote. Their answer in February 1912 was to strike. Over 1 million miners stopped work. The dispute lasted six weeks, before the government stepped in and promised a minimum wage.

That same year saw Mr Garforth re-elected as President of the National Institute of Mining Engineers. In his presidential speech, he pointed out that it only cost 0.01 d per ton of coal for stone-dust to be repeatedly spread throughout the West Riding Colliery. They had not had more explosions.

The universities that were established in the industrial areas of the country were venues for many mining engineers' meetings. It was these universities which were the first to honour William Garforth. The University of Birmingham bestowed a doctorate on him for his services to mining, followed closely by the University of Leeds.

In 1912, eighty-eight men were killed in two coal-dust explosions at the Cadeby Main Colliery, Yorkshire.

At 6 a.m. on the 1 February 1913 at the Lodge Mill Colliery at Lepton, near Huddersfield, James English and Albert Sykes rode down the shaft to carry out an inspection of a disused part of the mine, about a mile from the shaft. When they did not return at the appointed time, Albert Schofield, with another man, went to look for them. Albert was overcome; the other man managed to stagger back and give the alarm. The roadways were filled with gas.

The message was received at the West Riding Colliery at 9.30 a.m and the alarm sounded. Once again, for a moment, the village froze. The team on standby moved quickly, kissed their wives, collected their work clothes and ran to the colliery. In less than thirty minutes, the team was ready, but unfortunately a motor car was not available. It was 11 a.m before Mr Lloyd and three men were able to leave, and 11.30 a.m before the rest of the party left in a motorised charabanc, which broke down after three miles.

'Ready Bill?' Bill Burr looked up from the underground roadways map of the Lodge Mill Colliery. Every bone in his body ached. He stood, arched his back and looked at the rolling countryside. The ride over the bumpy potholed road had seemed endless. Mud was splattered all over him from the times they had climbed out to push the car. He brushed some off his work shirt and checked the adjustment of the leather straps around his chest once again. Then he looked across at Samuel Berry and William Webster, who had already donned their rescue apparatus.

'Yes Mr Lloyd.' Bill answered as he felt the weight settle on his back, then thought, 'I will be 45 in a few days. I'm getting too old for this.'

'Mixture okay Bill?' he breathed in as Mr Lloyd went through the now familiar checks to ensure his cylinders were delivering the correct mixture to his face mask.

'Time is five minutes past one o'clock.' Mr Lloyd announced, and each checked their watches.

The eighteen miles of road from Altofts had taken nearly three hours, raising the fear that they would once again be too late.

Each man signalled with a raised thumb, removed their masks and passed through a small group of men to enter the cage. At the shaft bottom, with sweat already running down his face, Bill checked his pocket watch. It was only ten minutes since they had arrived at the colliery. Seconds later, as the bang of the air doors subsided, quiet descended. A check was carried out for gas. Clogs stirred the mixture of fine coal and coal-dust, or thudded against wood sleepers and iron rails as each man, occupied with their own thoughts, moved forward.

They bowed their heads and shoulders to one side as they moved into a side road, where an area of roadway had been widened to allow for two sets of rail tracks.

'Good, the flat trams were there as we had been told,' thought Bill, checking for gas and placing his oil-flame safety lamp by his left side as he knelt on the tram. On his right, he placed one of the new electric lights they had been testing recently. Straightening his back, his head was just clear of the roof. He heard the scrape of metal on metal as Samuel Berry used his hands to push himself forwards. Bill waited for a short while, then placing his toughened hands on the small coal covering the roadway floor, he pushed. Keeping watch on the outline of Sam's body in front and a stopping distance between them, he pushed his way along. Only the swish of the rolling wheels disturbed the silence. Sam stopped and Bill moved slowly up behind him, adjusted the leather straps covering his knees, then dropped off his tram and crawled up beside him. The floor had risen, and the roadway height had been reduced. Wooden supports creaked as more pressure was exerted by the rocks above. With not enough room to lie on the trams with their apparatus on, and not willing to take them off, one by one they knelt behind their trams and pushed them forward till they had sufficient height to resume their place. A wooden prop cracked as they tested for gas, and after checking each others' apparatus, they moved on. Shortly after, Sam stopped again. This time, when Bill made to move forward, he was signalled back.

'Now what' thought Bill. Sam moved forward off his tram and disappeared. After a short while he returned, though he was now facing Bill.

'Side blow,' thought Bill, as Sam signalled him to turn the tram on its side. Careful not to dislodge large lumps hanging from the roof, with Sam pulling and Bill pushing, they eased the tram through. Where the pressure from the rocks on either side had built up, the width of the roadway had been reduced. Working as a team, they squeezed each tram through the stretch of narrowed roadway. Overcoming each problem and continually checking each others' apparatus for damage, they continued. Just after passing through a stretch of water, as they stopped again while rocks plopped into the water, a familiar blue haze showed its presence.

'Gas!' shouted Bill. All stopped and went through the process of fitting their mask to their mouths and noses. After once again checking each others' pipes and valves, they turned the cylinders on and moved off. Once again, Sam came to a halt and signalled Bill to wait as he moved forward. As Bill waited, his safety lamp flickered then went out. Fumbling in the total darkness, he switched on his electric light, trying to relight his lamp. There was so little oxygen in the gas filled atmosphere that it would not. Sam meanwhile had returned, and using the recognised signals with the electric light, informed Bill that they had to leave the trams.

Bill again adjusted the strips of leather covering his knees and crawled forward with the lamp held in front of him. Sam had moved into an older roadway off the main track and stopped. Bill pictured the underground roadway map he had studied, checked his watch, then crawled after Sam.

Supports stood at different angles. Large lumps had fallen or were half supported; fine rock had trickled down the walls. Conscious of the damage the splinters sticking out from the wood could do to their rubber pipes, they moved in. It was 300 yards before they found the first body. Bill once again checked his watch, 2 p.m.

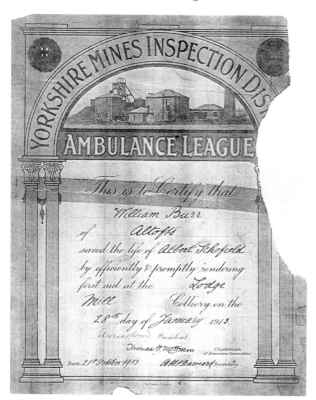

Certificate presented to William Burr after the rescue at Lodge Mill Colliery near Huddersfield.

Sam placed his hand on the body and signalled that it still had a pulse. With William in front and Bill behind, they carefully, inch by inch, supported the body, shuffling back to the trams. Placing it on the rear one, they pushed it backwards, taking extra care not to dislodge any supports. Time was precious. Bill took a deep breath and removing his face mask, held it to the mouth and nose of the man. Sam signalled no, so he replaced it and on they travelled. At each obstacle, they took the body off the tram, and then carefully manoeuvred it and the tram until they could continue. After each confined space, they checked each others' apparatus; one small tear could mean death. Eventually they reached a spot where they could relight their safety lamps and checking for gas, it showed clear. They left the tram and body and returned to where they had found it.

Sam, meanwhile, had found two more bodies, one of whom also appeared to still be alive. He had dragged them as far as he could. Now all of their training was beginning to pay, because normal men would have been exhausted, but they still had a lot to do. Once again they moved the body to a place with breathable air before returning for the last one. Once all three were in good air, Bill removed his face mask, did his gas check and took up position behind the first tram, the others doing likewise. Drenched in sweat, yard by yard, they crawled forward pushing the trams. After several rest stops, eventually, after almost a mile, they saw lights ahead

Bill again checked his watch. Quarter past five. He staggered to his feet as willing arms moved in to help him and take care of the bodies. Only his pride kept him walking, and fifteen minutes later, he tasted the clean, cold, evening air.

More output was now required, and the colliery continued to expand. The smaller coal was washed for local delivery.

New coal seams were required as older ones became worked out; these needed new headgears.

Local coal was now delivered by lorries. Note the trailers.

Unfortunately, the ups and downs of the coal trade continued.

These photographs show colliery staff keeping the boilers going.

By keeping the boilers going during disputes, this ensured men and management could immediately, after settlement of the problem, resume work underground.

Alfred Sykes was pronounced dead on arrival at the hospital. James English died three days later. Albert Schofield survived. Later he and Bill Burr, the man who had pushed and carried him through that hell, became firm friends. When Albert Schofield made one of his regular trips to his friend, the men of Altofts rejoiced with him. They had proved to the world for the first time that trained men could save lives, even under those terrible conditions. All those years of training in their own spare time had proved their worth.

As for Dr William Garforth and his wife Mary, the time, effort and money spent training and equipping the rescue teams had now resulted in the first life being saved after an underground explosion. It had been accomplished by his men, who had trained in the Mines Rescue School of his design, the first in the world. They had been wearing his rescue apparatus, also of his own design.

In 1913, as the Suffragettes were demanding better wages for the working girls, Dr William Garforth was once again voted President of the National Institute, for an unprecedented third time. He used the moment to suggest to the Home Office that the time had arrived when practical tests should be made to develop a more portable breathing apparatus, which each man could wear to use when carbon monoxide was present. However, it was to be more than thirty years before that became a reality.

It was also in this year that a Mines Rescue Station was built in Wakefield by the West Yorkshire Colliery owners to serve their colliers. Dr William Garforth offered the services of his men until the station was fully operational. He also later presented a shield so that the various rescue stations could compete against each other, thereby raising their standards.

The Rescue Team alarm shouted one last time in Altofts, and men, women and children stood and listened. It signalled the end of a remarkable period when the small village had been the focus of people worldwide.

However the campaign never ceased for Dr Garforth. Within a week, he was lecturing to the mining world again, trying to persuade them to use stone-dust spread on the collieries' underground roadways to prevent explosions.

The annual outing that year was a jolly affair, with Dr Garforth giving the families of over 2,500 West Riding colliers 2s each.

The following week, he addressed a meeting of several local town councillors. They had been arguing over the cost and the funding for a combined hospital. William stressed that even if they could save only one child's life, they must have a hospital. The arguments were resolved and the building work started.

In 1913, 440 men were killed by an explosion of coal-dust at the Universal Colliery, Senghenydd, near Caerphilly in Glamorganshire.

From 1913, all coal mines regularly employing more than 100 men had to employ men trained in Mines Rescue skills.

Those early members of the world's first trained Mines Rescue workers were: W.D. Lloyd, B. Harrison, J. Berry, W. Glegg, A. Jackson, J. Hopwood, J. Sykes, U. Holland, A. Whittingham, J. Jackson, T. Tredgold, B. Kellet, T. Wright, S. Berry, W. Newsome, J. Whittle, A. Tilson, S. Lucas, J. Brown, J. Toon, A. Bubb, S. Kitchener, J. Whittingham, W. Webster, W. Allbrighton, W. Burr, M. Watson, T. Newsman, J. Clamp, W. Hartley, I. Shackleton, S. Plimmer, A. Firth, W.W. Lunn, J. Cranswick, J. Bland, and C. Haslop.

fourteen

1914-21
RECOGNITION

The war clouds were gathering over Europe as Dr William E Garforth received a knighthood for services to mining in the New Years Honours list of 1914. The area celebrated with him. His family's work to improve the health and welfare of the community was showing results with the infant death rate being dramatically reduced. All houses had fresh running water and sewerage facilities. Men, women and children could improve their education at the local schools and colleges, at day or evening classes. Plans had also been made for a park and swimming pool, and Sir William was invited to become President of the newly opened Normanton Golf course. The week following his knighthood, he made a point of walking through Altofts, and any young person who called him Sir was rewarded with a sixpence.

As retiring President of the Institute of Mining Engineers, he was presented with a special medal 'in recognition of the eminent services he had rendered to mining'.

The nation's call for young men to enlist was echoed by Sir William. He encouraged all men under the age of forty to answer Kitchener's call to volunteer, and he gave a bonus of £2 to every man who did so. However, he understood the hardships the families at home would have to endure, so he promised each volunteer that his wife would receive 10s a week, and an extra 2s for each child. In just three days, 860 men volunteered, and when he saw them practicing drills with pieces of wood, he purchased rifles for them.

Many men from the West Riding Colliery joined the King's Own Yorkshire Light

Infantry (KOYLI), and along with other Normanton and Altofts men, formed their own companies. Others were a welcome addition to the various army units they joined. Very few non-medical men had first aid training, and less than 100 in the millions of men on both sides had Mines Rescue training.

The skills of the local miners were soon realised as the war developed into trench warfare, and mining and counter-mining of the opposition's trenches started. Here, the skills of the trained Mines Rescue workers were invaluable. However, the conditions they worked under, and the scenes they saw, were so horrific that none ever spoke of it after the war.

Sir William Edward Garforth

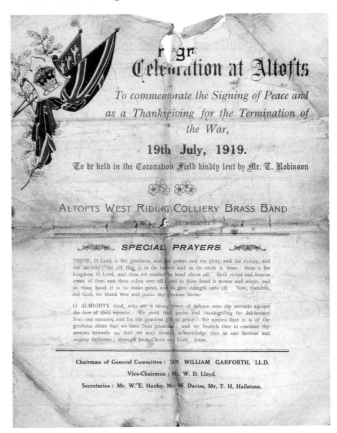

Celebration at Altofts

To commemorate the Signing of Peace and as a Thanksgiving for the Termination of the War,

19th July, 1919.

To be held in the Coronation Field kindly lent by Mr. T. Robinson

ALTOFTS WEST RIDING COLLIERY BRASS BAND

SPECIAL PRAYERS.

THINE, O Lord, is the greatness, and the power, and the glory, and the victory, and the majesty : for all that is in the heaven and in the earth is thine ; thine is the kingdom, O Lord, and thou art exalted as head above all. Both riches and honour come of thee, and thou rulest over all ; and in thine hand is power and might, and in thine hand it is to make great, and to give strength unto all. Now, therefore, our God, we thank thee and praise thy glorious Name.

O ALMIGHTY God, who art a strong tower of defence unto thy servants against the face of their enemies ; We yield thee praise and thanksgiving for deliverance from our enemies, and for thy gracious gift of peace : We confess that it is of thy goodness alone that we have been preserved ; and we beseech thee to continue thy mercies towards us, that we may always acknowledge thee as our Saviour and mighty Deliverer ; through Jesus Christ our Lord. Amen.

Chairman of General Committee : SIR WILLIAM GARFORTH, LL.D.

Vice-Chairman : Mr. W. D. Lloyd.

Secretaries : Mr. W. E. Hanby, Mr. W. Davies, Mr. T. H. Hailstone.

Sir W.E. Garforth's village celebrated the end of the war.

Bill Burr, because of his age, was not in one of the special units. He made few comments to his family and friends after the war, but from the army records, and the items and documents he left, I have pieced together some of his wartime service (see Appendix 2). People can only imagine the experiences of the ones employed full-time in this role.

The war carried on, and many local men were awarded high military decorations, (see author's book 'Normanton's Great War') but hundreds were killed and thousands wounded. Some came to the local hospitals in Yorkshire; some were sent to hospitals all over Europe. In one of these, Stephanie Garforth was doing voluntary work.

'What do you look like, Miss Garforth?' asked one blind soldier who had been gassed, between coughs, as Stephanie led him around the hospital gardens.

'She is beautiful.' answered another, a curly-headed youngster with just stumps for legs.

'I wish she would hold my hand,' joked another, who had a bad chest wound.

'Well I do need a partner for the hospital dance this weekend,' answered Stephanie who, after two weeks, was able to control her face so no sadness showed.

'I will!' chorused men on crutches and others in wheelchairs as they surrounded her.

All her life, Stephanie and her sisters had helped her mother Mary in the local community around their home, but nothing could have prepared her for this.

'You must get some rest,' urged Mary as Stephanie returned home after another long stint.

'If you could only see them,' Stephanie answered.

His monument.

The following day, the same blind soldier, sensing the unusual atmosphere around him, asked,

'What's happened?'

'You know that beautiful young woman who made us all jealous by holding your hand yesterday?' a patient replied quietly.

'She was running to help someone who had fallen on the floor when she slipped and banged her head against a wall. She's dead.'

At the end of the war the local community counted the cost. Most of their men had served in the Kings Own Yorkshire Light Infantry (KOYLI) Battalions. Their operation strengths were approximately 1,000 per Battalion. As casualties occurred, newly recruited local men made the numbers back up to 1,000. The First and Fourth Battalion KOYLI lost forty-eight officers and 630 men killed, with a further 1,560 missing, presumed dead. For the Second Battalion KOYLI, out of the original 1,000 men who had started the war, only one officer and thirty-four other ranks survived.

Sir W.E Garforth's grave.

When Major-General Douglas Haigh, the commander-in-chief, passed through Altofts by train, he bowed fifteen times as they cheered.

In April 1921, Sir William was invited to become chairman of the Altofts council for the forty-first time. He declined, saying, 'Kindly excuse me on this occasion.'

In July, the government, after calls from Sir William and many other distinguished mining engineers, set up the Mining Dangers Board, to look into all aspects of the safety and welfare of the mines and miners.

Later the same year, it became law that stone-dust must be regularly spread throughout coal mines. William's persistence and endeavours had succeeded at last.

On 1 October 1921, 'one of England's noblest gentlemen' died. Many thousands paid their respects as Sir William's cortège slowly moved from Snydale, via Normanton church, to the Altofts graveyard.

APPENDIX 1

William and Mary Garforth were still recalled with grateful words by many people when I started my research in the early 1980s. Mary died on the 1 September 1941, and is buried with her husband, along with daughters, Stephanie, Francis, Sarah, and Mary Eager. Only two of his daughters were to marry: Katherine who had two sons and a daughter, and the eldest, Margaret, who had a daughter.

On a further note, William was also Grand Master in the Freemasons. Their principles are: brotherly love, relief and truth.

Coal production at the West Riding Colliery ceased in 1962. One final point I would like to mention is that George Allat's burnt Bible (mentioned on page 24) is now in my posession.

My sincere thanks for their help in providing access to the records go to:
Mrs M Richardson, of the former Whitwood Mining Technical College. Incidentally, the main instigator for the foundation of this college was William Garforth.

Mr M Ryder of British Coal Records, The City of Manchester Local History Library, the Tameside Local Studies Library at Stalybridge, and Wakefield Local History Library.

Also, the many people who helped to collate this book.

Finally, once again, thanks to the people of Altofts and Normanton, whose memories made this book possible.

APPENDIX 2

Let us look a little closer at one of Mr Garforth's miners.

Born in Maidstone in 1868, William Burrluck had, as a young lad, walked to Wales, but then, hearing there was a better chance of work in Yorkshire, he had continued to Altofts. 'I've left my bad luck in Maidstone,' he had told his friends, as he dropped the last letters of his surname and was now known as William Burr. Bill, like most of the men, could not read or write. Once in Altofts, he, like many others, lodged with a family, the Schofields. Their eldest daughter, Harriet Henrietta, known as Kit, taught him to read and write. He formed, and captained, the West Riding Colliery football team.

On 3 August 1889, William Burr and Harriet Henrietta Schofield married. In 1892, their first daughter, Jane, was born. In 1896, Lucy was born. Also, Bill's friends arranged a social evening and presented him with a framed photograph of the football team he had established in the village.

In 1898, when daughter Hanna was born, her parents despaired of the future. Bill had been on strike for over six months, they had no money and very little pride left. They gratefully accepted any help. Their family eventually grew to include seven daughters and two sons.

1906 was an example of one of the happy times. A happy crowd of families and employees of the West Riding Colliery started from Altofts, growing larger as it passed through Normanton, then they walked the three miles to Snydale Hall, William Garforth's new home, for the garden party that the family were hosting.

Jane Burr, now fourteen years old, occasionally carried baby Alice, to give Harriet a rest. Hanna, seven, and Lucy, nine, played and ran with the other children. At one stage, the whole dusty column, now numbering in the thousands, stopped and waved as a motor vehicle spluttered its way through them.

'Dad,' Jane said, 'have you won a giant silver cup for first aid?'

'I was one of a team,' Bill replied, 'just like us,' he put his arms around his wife and eldest daughter. They talked about garden party for weeks afterwards.

This is Bill Burr's story between 1914 and 1918, but it is a tribute to the millions of men who went to war. A beautiful day dawned as Bill Burr stretched out lazily on the luxurious growth of grass. The air tasted clean and sharp. Overhead, fluffy white clouds chased each other over an azure blue sky. Birds were singing and he could hear the rippling sound of water in the nearby river rushing downstream after yesterday's rain. He opened Harriet's letter once again, which had included notes from all his daughters. They could all read and write; he was so proud. He looked at the photograph she had sent of their first son after six daughters.

'I've hardly seen him,' he said aloud. John Thomas, or Jack, had been born on New Years Eve, 1914. Bill had volunteered on 20 August, almost the first day of the war, and he had seen little of his family in the last couple of years. He re-read Harriet's letter. Poor Lucy, his second-eldest, had recently celebrated her twenty-first birthday and was already a war widow. He neatly refolded the letter. He closed his eyes and took a deep breath, allowing the picture of his family to come into his mind.

A distant rumble and flashes of light made him aware of the passage of time. A sudden enormous flash of light, then waves of sound, beat against his ears; the vibration from the ground tingled along his back and as he looked to the east, a dark grey cloud filled the horizon. 'Time to go!' came the shout.

He picked up his cape and pack and joined others as they quickly ran for the transport. Climbing up on a horse-drawn waggon, the gaunt horses, despite their fresh feed of grass, did not appear to have the strength to pull it. He waved gaily as they passed young farm girls working in the fields and they waved back. As the mud deepened, they climbed off to help the tired horses. At a small village they stopped. Instructions were given and more equipment was thrown on the wagon, and once again they helped the horses pull the wheels through the deep mud.

'Everybody off!' the driver shouted back as they stopped again. Wiping the mud from his hands, he passed the packs off the wagon

When his number was called, Bill shouted, 'Here.' and a large pack was thrown to him. They all stood and listened to a quick series of instructions then, slinging the large pack on his back and hanging another over his shoulder, Bill set off.

Abruptly, the landscape changed. All around stood trees which had been shorn of their leaves and branches. No grass showed through the deep mud, and he heard what sounded like thousands of hailstones on a tin roof. He smelled and tasted the lime blown around by a gentle wind as he scrambled into a deep trench with wood supported sides. The wooden catwalk groaned as it settled onto the supports driven into the deep mud. His mind became blank as all his efforts were directed towards progressing in the now almost-knee deep mud to his destination.

Occasionally they stopped to allow the movement of their heaving chests to settle. As the noise of their breathing slowed, they heard the rattle of machine guns and the occasional pop and blast of heavy shells dropping all around. As they neared the front, the mud got deeper and the smell of rotten flesh stronger. He did not flinch as he disturbed an enormous rat, feasting on the last vestiges of flesh on a skeleton which had been thrown into the trench by the last round of shelling.

Coarse jokes about the Huns were exchanged with men who perched on the drier spots, while an officer used a periscope to peer over the side of the trench. They passed others huddled in shelters dug into

William Burr.

the walls of trenches, smoking or reading letters from loved ones. Braver men even tried to snatch a quick bite to eat while doing battle with the swarms of flies. Occasionally they passed a deep dug-out where snores indicated men were snatching a few hours peace, away from the environment of war, nestling among fat rats to keep them warm, hands clasping their gas masks as a precaution against creeping death.

Bill's progress along the trench was abruptly halted by a wall of mud. They had reached their destination. The Germans had managed, without detection, to dig under a strong point, and the great flash of light and wall of sound had been their mine exploding.

'First aid kit!' Bill knew the drill as he handed back his first aid pack and Red Cross armband; they were of no use covered in mud and he called up fresh energy as the desperate race against time began.

'My turn first,' he said as he quickly tied on his knee pads for protection against the barbed wire that the explosion had blasted into the soil. With shovel in hand, he dropped to his knees and pulled his eye shield into place. The rhythm of shovel-and-throw, honed to perfection on the coal faces of the West Riding Colliery, was quickly established. Behind him his partner, known as Taffy, ensured that most of the mud thrown back by Bill went over the trench side.

'My turn for the easy life,' Taffy said after half an hour, and they changed positions, Different muscles now took the strain. Most men could not have laboured more than a short period of time, but with hands of iron and forearms out of proportion to the rest of their bodies, they removed the mud and slime as quickly and efficiently as any digging machine. Only the lengths of barbed wire that were uncovered caused a short halt to the procedure. Bill, ever alert for a different noise or feel as the shovel sank into the mud, stopped, his muddy fingers uncovered a skeleton. Picking up his shovel, he moved it. Time was precious in the race to save more of their comrades from joining it.

'Second XI's turn,' said Bill, thinking back to his football days. Mud splashed over the side of the trench as a shell fell nearby. The second pair of former miners took their places, the rhythm never slackening. The Germans, alerted by the activity, constantly sprayed the area with machine guns. A uniform was unearthed, and hands replaced shovels, but they were too late. The body was passed to Bill and he passed it down the trench to be piled on top of others on a stretch of dry walkway.

Bill and his mate returned to strengthening the trench side with timber, to prevent it collapsing. Others had learned the hard way the price of disregarding their own safety. While others dug, they did battle against the muddy water seeping in from all directions. Another pair crawled through the water-filled shell holes, looking for wounded or dead men outside the trench. Other team members sorted out the many body parts like jigsaws and placed them in sacks, whilst similar teams worked in nearby trenches.

'Dug-out!' was shouted back, and while the entrance was uncovered, wood was passed forward. With aching muscles, they cleared the mud and tore open the door, releasing stale air and mud. Efforts were redoubled as they found the hollow, hopes raised. Once a space had been cleared, Bill and his mate moved forward. The men at the entrance staggered drunkenly to their feet to allow them to pass. Standing, with muddy hands they carefully pulled clean pieces of cloth from their pockets and wiped their eye shields. Meanwhile, wood had been quickly passed down the trench and the entrance of the deep dug-out was being shored up. Others explored the hollow as Bill and Taffy resumed their rhythm, and in two hours, they uncovered 20 yards of former trench, and four more bodies.

'Another dug-out!' Bill called back as his shovel slid into a deep hole. As he turned back he stopped and stiffened. From his vast experience, he knew that smell. 'Gas!' he shouted as he stood up to his waist in mud. As he heard the message being passed back, he pulled on his gas mask. The gas from the gas-shell flowing past him would disperse in the open air, but he needed his rescue apparatus on in order to explore the dug-out. The team knew the procedure. As he waited, his mind went over the many different Mines Rescue operations in which he had taken part while a member of the colliery rescues team.

He remembered the sunny day when almost the whole of the football team had walked happily to Pontefract.

'You're first aid trained?' queried the recruiting sergeant, looking at Bill's filled-in form, 'We shouldn't have much use for you; it'll all be over by Christmas. But if you're good with bandages, the Medical Corps will be the best regiment for you.' Once his training had been completed, Bill had been attached to the Guards.

Many times in the months that followed, Bill had crawled out into No-Mans Land, between the trenches, to use his skills on the wounded of both armies.
On a rainy day in 1915, when the regiment had been on parade, the Guards Division Commander, Major General E. Fielding had paid a visit. 'Regiment, Attention!' The commanding officer shouted, and then proceeded to call out names, all of whom went forward to be decorated with medals for valour.

'Medical Orderly, William Burr.' Amazed, Bill took a short while to react to the summons, then marched smartly forward and saluted. 'This presentation is to recognise the number of lives this soldier has saved by his work of establishing forward first aid posts, and his keenness and resourcefulness in applying first aid to soldiers under fire.' an officer read out.

'You are to be congratulated on your devotion to duty,' said the General as he presented Bill with his certificate.

'We have some miners here,' the commanding officer addressed the unit after the General had left, 'Have any of you any knowledge of Mines Rescue?'

'Yes Sir,' shouted Bill, who was then beckoned forward. No other soldier answered the question

'Right,' the commanding officer addressed Bill, 'you will train a special section, under an officer I will appoint, to use that,' and pointed to a large box. 'Now, I want some volunteers to be trained,' the officer shouted then called out the names of his volunteers.

Bill went over to the box, opened it and began removing the packaging until he came upon a document. As his volunteers gathered around, he picked up the instructions and began to read. It was a list of people who had been involved in the design. It was then that he saw the name he knew so well,

'Sir William Garforth.' he read aloud. He straightened up and looked around, 'Sir William Garforth was my colliery's manager.' He paused to allow that to sink in, then continued to read out the instructions. 'Now to show you how to wear it.' Although the apparatus was slightly different to the WEG they had used at the West Riding Colliery, he soon had the equipment over his head and was demonstrating how it worked.

The team quickly learned how to use the apparatus. But the first time it was used in action, no-one had the same strength and stamina that he had developed, to wear it for long. He showed them the physical training routine he had gone through so many times.

'Not bad for a forty-eight year old,' he had said afterwards as he looked up at the giant guardsmen who formed his team, all of whom had been selected because they were also former miners.

Many times he had donned the apparatus, as the war had stagnated into trench warfare. Each side had tried to gain an advantage by digging under the opposing side's trenches. They then filled the resulted spaces with gunpowder and an occasional gas-shell, blowing up the trenches above. Sometimes the explosion went off early, catching both sides, and with the gas lingering in the remains of underground roadways, it made it impossible to work without rescue apparatus. Bill's first job was always to search for anyone alive, then to remove the dead bodies or parts of bodies. He remembered the number of times his searches had resulted in the dragging out of Germans, both dead and alive.

At a church parade the previous Sunday, the commanding officer had called, 'Today we have a very unusual request. We have been sent a German Iron Cross to present to a soldier who has saved many German lives, Medical Orderly William Burr.' Bill remembered his embarrassment as he had marched forward to collect it.

The rescue apparatus arriving disturbed his thoughts. That was the past; this was today. Aided by Taffy, he donned it, ensuring the eye-pieces and breathing tube were clear. Then he adjusted the valves for the correct oxygen mixture. He tied a rope around his waist, and giving the thumbs up, dropped to the trench floor and began enlarging the hole. Time was now limited by the mixture on his back. The hole was the top of a dug-out. Gas was lighter than air, so possibly there would be pure air at the bottom. A skeleton partially blocked the entrance, but as time was precious, it was quickly removed. Now he pushed his electric light forwards and moved further in. Wood was placed at his side to strengthen the roof and, placing a piece on the floor, he slid on the mud to the bottom of the dug-out. More wood quickly followed. His light showed where the roof required strengthening and he quickly hammered home roof supports.

Bill moved forward slowly, a rat overcome by gas toppling sideways into the mud. But he had lived up to that moment, so maybe there was a chance. The rat's dinner was dead; the body could be collected later. Another was directly in his way. He pulled in some more wood and attached the body to it. He signalled four pulls on the rope, and it was dragged up until hands could reach down and pull it from the mud.

Once again he looked around, a broken piece of a door sticking out of the dug-out's side showing were there had been another entrance. He slid across and began clearing the entrance. Occasionally stopping to pull in more wood to strengthen the roof, he signalled when the rope was ready for its backward journey. Each bundle of wood gathered its share of mud as it approached him. His pull had to be measured to ensure that the steady flow of oxygen mixture was not ruined by the racing of his heart.

'I'm getting old' he said to himself, the mud squelching as once more he lifted his legs from it. Now he could place the roof supports in position and after another couple of pushes with the shovel, the mud on the entrance slipped away. As Bill bent forward to push it to one side, his hand touched the cloth of a uniform.

'It must be an officer,' he thought, as he felt the smooth material. He shone his torch on the roof which had to be supported before he could progress forward. He pulled a small pouch out his pocket and extracted his watch. It was imperative he knew just how much time the cylinders on his back allowed him. The supports arrived and were hammered into place and he moved forward. The roof looked solid. Taking a chance, he moved quickly into the dug-out. As his light swept around the space, a body emerged out of the gloom.

Diamond News

No.78

80 years of Mining Machinery Manufacture!

1897–1977

The B57 In Web Shearer

A captain in the Guards, his face was above the mud. He moved across, wiped his hands and checked for signs of a life, feeling the faint flicker of a pulse. Quickly he looked around, spotting a table top. He dragged it next to the body and turned it over. Using the remainder of his energy, he pulled the body onto the table. His eye-pieces were now steaming up with the heat his body was creating as he tied the rope around the table and body, and gave four pulls.

He allowed his heart once again to slow down, then took his watch pouch out. He still had twenty-five minutes. He moved across to where a lack of side-supports indicated what must have been another entrance to the dugout. Using a bed and its boards, which had been piled in one corner, he started the task of clearing the entrance. Once again, he used minimum force to push the shovel into the mud. Slowly he cleared the doorway from the top down. His shovel hit something hard, so using his fingers he cleared away the material around the object and found a boot. Bed-boards used up, he braced himself against the side of the tunnel and pulled forward the wood that had now been attached to his rope. Watch pouch out, his breathing slowed down again. He had fifteen minutes, just enough time to get the body out. Slowly he uncovered the body, ensuring once again the roof above him was safe. He rolled it onto two wooden planks he had pulled in, and tied the rope around it and the planks. Then he made four tugs on the rope. Once again he took his pouch out. Time had run out and he staggered out, into the safety of the smell, taste and noise of war. The captain later presented Bill with a pocket watch, which the family still treasure.

Bill Burr, after he had returned from the First World War, discovered he had a remarkable gift. He could cure dislocations and sprains. His fame became legendary. Thousands of sportsmen and women visited him; he never charged a penny. His most treasured possession was a first aid casket presented to him after a public subscription. Even on his death bed in 1959, aged 91, he continued to use his remarkable powers. His seven daughters and two sons were, Jane, Lucy, Hanna, Alice, Dora, Harriet Henrietta, May, John Thomas and William. Harriet, his wife, died in March 1940.

His family supplied many of the photographs, and their memories.

Other local titles published by The History Press

Olde Yorkshire Punishments

HOWARD PEACH

Locals and visitors alike will be fascinated by this revealing insight into the dark world of Yorkshire punishments through the centuries. This lavishly illustrated book covers all forms of punishment, ranging from the gruesome and extreme to the downright peculiar, such as the Church's 'dog-whippers'. It will appeal to all those wishing to discover more about Yorkshire's intriguing past.

978 0 7524 4661 5

Paranormal West Yorkshire

ANDY OWENS

Poltergeists. UFOs. Murder mysteries. Big cats. Cases of human combustion. Victorian *causes célébres*. With famous cases such as the Cottingley Fairies - investigated by Sir Arthur Conan Doyle - now forgotten cases such as the Pontefract Poltergeist, this richly illustrated collection covers a fascinating range of strange events from West Yorkshire's history. Including sources both ancient and modern and with never-before published investigations by the Haunted Yorkshire Psychical Research Group, this book will delight all lovers of the unexplained.

978 07524 4810 7

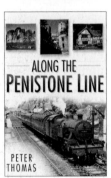

Along the Penistone Line

PETER THOMAS

It is curious that the nineteenth-century railway builders should even have thought of a line linking Huddersfield with Sheffield, now known as the Penistone Line, with the risk of all sorts of disasters. As *Along the Penistone Line* makes clear: collisions, runaway trains, collapsed tunnels and viaducts, the Penistone Line had them all. The operating costs of The Penistone Line made it a prime target for closure time and time again, yet against all the odds it survived and has become the heart of a community.

978 07509 4619 3

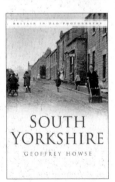

Britain in Old Photographs: South Yorkshire

GEOFFREY HOWSE

South Yorkshire in Old Photographs provides an illuminating insight into the history of this fascinating area, which includes Sheffield, Rotherham, Doncaster and Barnsley, as well as innumerable smaller towns, villages and hamlets. These carefully chosen photographs depict what life was like in bygone days, from beautiful village scenes, and stately homes, to now-vanished industries, houses and streets, along with perhaps the most important aspect of South Yorkshire – its people.

978 0 7509 4658 2

If you are interested in purchasing other books published by The History Press, or in case you have difficulty finding any History Press books in your local bookshop, you can also place orders directly through our website
www.thehistorypress.co.uk